ULTIMATE ENGINEERING

ULTIMATE ENGINEERING

AN ENGINEER INVESTIGATES THE BIOMECHANICS OF THE HUMAN BODY

STUART BURGESS

SEATTLE   DISCOVERY INSTITUTE PRESS   2026

Description

Who first invented fiber optic cables? Or camera shutters? Or truss structures? Or double wishbone suspensions? Or block-style Roman arches? What about laser beams? As award-winning British engineer and designer Stuart Burgess reveals, the original inventor of these and countless other ingenious devices was no human inventor. Instead, the "first to market" was the designing force behind the living world.

Evolutionary theory predicts a living world crowded with substandard designs. But as Burgess shows in *Ultimate Engineering*, the latest science has discovered just the opposite—designs so advanced they are at the limit of the possible, precisely as proponents of the theory of intelligent design anticipated. As Burgess also details, he and other researchers are taking the discovery of these advanced designs and using them to inspire fresh technological breakthroughs—a revolution known as biomimetics.

Library Cataloging Data

Ultimate Engineering: An Engineer Investigates the Biomechanics of the Human Body
by Stuart Burgess
Cover design by Tri Widyatmaka.
374 pages, 6 x 9 inches
Library of Congress Control Number: 2025948239
ISBN: 978-1-63712-084-2 (paperback), 978-1-63712-086-6 (Kindle), 978-1-63712-085-9 (EPUB)
BISAC: SCI036000 Life Sciences / Human Anatomy & Physiology
BISAC: SCI027000 Life Sciences / Evolution
BISAC: TEC059000 Biomedical

Publisher Information

Discovery Institute Press, 506 2nd Avenue, Suite 1700, Seattle, WA 98104
Internet: discovery.press
Published in the United States of America on acid-free paper.
First Edition, February 2026

ADVANCE PRAISE

I am very pleased that Stuart Burgess has produced this excellent new book, *Ultimate Engineering*. From his extensive experience in various aspects of engineering design, including biomimetics, Dr. Burgess is well qualified to write about design in nature. As I have said before, the evidence for intelligent design invalidates any theory of evolution. Burgess's many examples of "ultimate engineering" in these pages powerfully reinforce that conclusion.

From my lifetime of researching bacteriology, I know that complex structures in the plant and animal kingdom cannot be produced by thousands of minute evolutionary steps of ever increasing complexity over millions of years. Darwinism is a man-made theory to explain the origin and continuance of life on this planet without reference to a designer. But why rule out intelligent design? As I have also said before, any properly complete theory of man's origins must embrace the whole man, including feelings, emotions, pleasure, beauty, morals, motives, final causes, and purpose.

It is my hope that this book will lead its readers to see that there is evidence in nature not just of intelligent design, but of a great and deeply purposive intelligent design, especially in the design of humankind.

—**Alan Linton**, PhD, DSc, Emeritus Professor of Bacteriology, former Head of the Department of Microbiology, Bristol University, UK, Honorary Associate of the Royal College of Veterinary Surgeons, Fellow of the Royal College of Pathologists

It is a privilege to commend this superb book. Any rational response to Stuart Burgess's case for intelligent design, laid out in these pages, should include wonder and an acknowledgment that there is an awesome designer responsible for our universe. My professor of physiology at medical school in 1966, in his first lecture, told us that we should never describe what some part or process in the human body was for because that implied there was a designer. That same battle of worldviews exists today.

In the first fifteen chapters, Burgess shows that "biological systems are engineering systems." The final seven chapters draw attention to the realities that stack up against evolutionary theory. Burgess shows that Darwinian natural selection and mutations cannot produce the wonders seen in the human body. The arguments and confessions of evolutionists are also well documented. One senior professor, a non-religious microbiologist, suggested that evolutionary theory was "black magic." Another such professor, a leading biology researcher at a top university, said in an equally frank moment, "You wave a magic wand and say, 'Evolution did it.'"

Throughout the book, Burgess answers all the claims for bad design one by one. Time spent reading *Ultimate Engineering* could lead to a new appreciation of where you came from, why you are here, and where you are going.

—**Nigel Jones**, MS, Fellow of the Royal College of Surgeons of England

Ultimate Engineering sets out a fascinating array of brilliantly engineered features in the human body at the visible anatomical level. Along the way, the book faces head-on the claim that the body bears the hallmarks of poor design, the result of unguided evolution by natural selection. The supposed "backward wiring" of the human eye, far from being a design weakness, as evolutionists claim, is actually a design strength. The purportedly unnecessary bones of the foot-ankle complex instead contain "an ingenious triple-arched structure that has astounded engineers." The knee joint, ostensibly the substandard

result of evolutionary tinkering, is actually a textbook example of a "four-bar linkage mechanism." These are just a few of the examples included in this masterclass on the extraordinary engineering feats in the human body, designs that have professional design engineers in awe. Professor Stuart Burgess, an award-winning design engineer with an outstanding academic and professional reputation, explores all this with great clarity and with the authority of a world-class biomimetics inventor and researcher. All students and teachers of STEM subjects interested in origins and biomimetics should read this book.
—**Stephen Palmer**, MA, FFPH, former Mansel Talbot Professor of Epidemiology and Public Health, Cardiff University, Fellow of the Royal College of Physicians of London

This book has assembled a plethora of evidence for design in human biology and has done so in a most persuasive way. The approach is unmistakably that of a mechanical engineer, and the way Stuart Burgess has exposed those with the audacity to make unsubstantiated claims of "bad design" carries genuine academic credibility. The author's personal experience of faltering opposition to a design model is interesting and surely bears out the claim that a tipping point in scientific thinking is fast approaching.
—**David Galloway**, MD, former President, Royal College of Physicians and Surgeons of Glasgow, author of *Design Dissected*

Who else better to write this important book than Professor Stuart Burgess, the globally renowned engineering designer. His book is packed with extraordinary examples and insights. I thoroughly recommend it.
—**Colin Garner**, PhD, Emeritus Professor of Applied Thermodynamics, Fellow of the Royal Academy of Engineering

In *Ultimate Engineering*, leading UK engineer Stuart Burgess culminates a career of rigorous analysis of life's amazing and coherent engineered systems. His easily accessible writing wraps a brilliant and astute core insight that is *not to be missed*—which he neatly sums up in Figure 19.5. Don't hesitate to give this diagram a sneak peek as you begin your journey through this wonderful book.

—**Steve Laufmann**, systems engineer, chair of the program committee for the 2023 Conference on Engineering in Living Systems, and co-author of *Your Designed Body*

Most great men I have met are humble but brave. Professor Stuart Burgess is one of them. *Ultimate Engineering* is a proper name for this great book, in which Stuart shows that the engineering of structures and processes in the human body is superior to any achievements of human engineering. Intelligent design is an obvious explanation for all those who are not blinded by the evolution paradigm, which has its roots in Hegelian process philosophy. Engineering researchers stand firmly with both feet in reality. They rely on observations and logic, and quickly understand the limits of evolutionary processes.

—**Matti Leisola**, DSc, bioengineer, former Dean of Chemistry and Material Sciences at Helsinki University of Technology, winner of the Latsis Prize-ETH Zürich, and co-author of *Heretic: One Scientist's Journey from Darwin to Design*

What happens when an engineer—one who has spent a lifetime designing elegant complex systems—takes on evolutionary biologists who deny that living systems were designed? The answer is this brilliant book. Stuart Burgess knows design when he sees it, because he has decades of experience designing things. His engineering credentials are impeccable, and they allow him to enthrall us with insight after insight into why life works the way it does. Spoiler alert: It's not because life was cobbled together through blind evolution. It's because the human body—literally from head to toe—was designed by a masterful designer.

—**Casey Luskin**, PhD, co-author of *Science and Human Origins*, *Traipsing Into Evolution: Intelligent Design and the Kitzmiller v. Dover Decision*, and *Discovering Intelligent Design*

If you feel the discussion of origins has become bogged down in the conceptual mire, *Ultimate Engineering*, by Stuart Burgess, is the book for you. By bringing his internationally recognized expertise in design and engineering to assess human skeletal anatomy and broader physiological processes, Professor Burgess demonstrates that the shallow, Darwinian logic that humans are either poorly designed or only just good enough to survive just doesn't fit the evidence. By presenting compelling examples of "over-engineering" in the structure and function of the human body, Burgess reveals that these are expected outcomes from a process of intelligent design—the masterpiece of God's creational prowess. The additional accounts of his own experiences advocating for intelligent design in the UK academic setting are illuminating, adding a personal touch that will surprise many.
—**Tim Wells**, BSc, PhD, Emeritus Professor of Neuroendocrinology, School of Biosciences, Cardiff University

Stuart Burgess has written what will likely become a defining book in the ongoing debate over whether life is the result of purposeful design or an unintended accident. Drawing on numerous examples, he demonstrates that living systems consistently exhibit engineering solutions that appear to represent the best designs physically possible. He further argues, based on the testimony of evolutionists themselves, that undirected processes must produce results far less efficient and elegant than those of human engineers, the opposite of what is observed. Burgess reports several cases where the denial of design has led biologists to conclude that a trait, such as the human ACL or foot-ankle complex, is flawed or even non-functional. Yet his own research has repeatedly shown the opposite. Moreover, Burgess credits his enormous success in his career to recognizing that life is designed and learning from its ingenious strategies. The book also offers an eye-opening personal narrative. Burgess recounts conversations with biologists and engineers who privately admit that the evolutionary framework does not fit the evidence they observe. Yet they feel the need to remain silent out of fear of the professional consequences of challenging the status quo. In summary, the book argues power-fully that a design-based perspective can free biological science from materialist restraints and accelerate scientific discovery.

—**Brian Miller**, PhD, complex systems physicist, Research Coordinator for Discovery Institute's Center for Science and Culture, and organizer of the biannual Conference on Engineering in Living Systems

If you really want to think outside the dark box of evolutionary theory, then Stuart Burgess's concept of ultimate engineering will definitely give you a light and a path in your quest for discovery. I can't emphasize enough how this book has deepened my thinking and raised arguments against the established views of evolutionists. I would strongly recommend *Ultimate Engineering* for anyone with rational thinking to read and enlighten themselves.

—**Muhammad Saleem**, MBBS, MRCS, trauma surgeon, United Lincolnshire Teaching Hospitals, Fellow of the Royal College of Surgeons of England

Finally! An easily accessible and superbly written book, from a seasoned professional with stellar credentials, that skewers the (supposedly) "bad design" arguments of those quack engineers, aka evolutionary biologists. Learn incredible things about how your body actually works that only a design engineer in biomimetics can teach you. Enjoy the ride and pass it on!

—**Howard Glicksman**, MD, co-author of *Your Designed Body*

In his book *Ultimate Engineering*, Stuart Burgess takes intelligent design arguments to a new level. Based on forty years of experience in construction engineering and biomimetics, and with more than two hundred scientific publications, Burgess does a lot more than argue that nature exhibits purposeful design. He also demonstrates very convincingly, with multiple examples, that this design reveals ultimate engineering and diversity. This finding strongly points towards a highly intelligent and creative Designer, and it is the opposite of what the evolutionary paradigm predicts. I would say that after reading this book with an open mind, it will be very difficult to be both a fully informed *and* an intellectually fulfilled Darwinist. I very much hope that *Ultimate Engineering* is widely read, and that many people who read the book want to get to know the Designer of nature.

—**Ola Hössjer**, Professor of Mathematical Statistics, Stockholm University, probability theorist with applications in population genetics, epidemiology, fine-tuning in biology, and the limits of evolutionary theory; winner of the Gustafsson Prize in Mathematics

Dedication

To the many academics and students who have supported my public stance on intelligent design and encouraged me to write this book. I have been involved in supervising more than thirty PhD students, and this has always been inspiring and a great pleasure. Many of the examples I give in this book have come from biomechanics research I have carried out with my postgraduate researchers. I am also thankful to the three universities I have worked for (Bristol, UK, Cambridge, UK, and Liberty, VA), including their support for my freedom to share the evidence I see for intelligent design.

CONTENTS

INTRODUCTION

Two Views of Biology

The more we learn about living systems, the more we find ourselves stunned by the masterful designs in biology. Andrea Rinaldi, an evolutionary biologist and world-leading fungi expert, whose papers have been cited more than 7,000 times, put it this way:

> Biologists often find themselves awestruck by the elegant perfection of living organisms.... From molecules to organisms, scientists and engineers have repeatedly been enthralled by nature's handiwork and have emulated natural designs in man-made innovations.[1]

I am one of the engineers he's describing. I have helped design prosthetics, spacecraft for the European Space Agency, and the fastest track bicycle in the world for the British Olympic team. For four decades I have worked alongside top researchers in biology and engineering, and together we are not only awestruck by the engineering marvels of the biological realm but inspired by them to make significant engineering breakthroughs outside of biology. That pursuit, now a subdiscipline of its own, is known as biomimetics.

Despite all this, some evolutionists, including Nathan Lents, Abby Hafer, Jerry Coyne, and Richard Dawkins, insist biology is characterized by bad design. They further argue that this supports the theory of evolution, since Darwin's mechanism of natural selection working on chance variations (now understood as random genetic mutations) is a mindless, trial-and-error process that can be expected to have routinely drifted into decidedly suboptimal design solutions in the history of life.

What's behind this dramatic divide among life scientists? On the one hand, we have experimental biologists, like Rinaldi, who are reporting detailed scientific observations from the laboratory that reveal ultimate engineering in biology. On the other hand, we have biologists such as Lents and Dawkins, who make claims of bad design based on the evolutionary prediction of bad design, and typically do so in areas of biology for which they lack expertise. This latter group are so keen to see suboptimal design in biology that they overlook evidence of ultimate engineering.

Evolutionary Theory Predicts Bad Design

Why do evolutionary theorists so often anticipate bad design? Because the evolutionary mechanism—as understood by both Darwin and by modern evolutionary theorists—is highly constrained. As Duke University professor of biology Steven Vogel put it, "The dazzling diversity of the living world too easily disguises the fact that the evolutionary process faces constraints far more severe than anything impeding human designers."[2]

The most severe constraint on evolution is that of incremental change, which prevents it from producing systems that require many parts to originate simultaneously. But there are other constraints. Evolution is constrained by a limited ability to shed vestigial parts, and it cannot evolve much beyond the organism's survival/reproduction needs. (These constraints will be further explored in Chapter 17.) Evolution is so constrained in producing changes that evolutionary biologists are well within their rights to conclude that it could only produce biological systems with many areas of poor design.

Lest one imagine I am presenting a false view of modern evolutionary theory, consider this passage from distinguished French biologist François Jacob, in his essay "Evolution and Tinkering":

> The action of natural selection has often been compared to that of an engineer. This, however, does not seem to be a suitable comparison... because the objects produced by the engineer, at least by the good engineer, approach the level of perfection made possible by

the technology of the time. In contrast, evolution is far from perfection. This is a point which was repeatedly stressed by Darwin who had to fight against the argument of perfect creation. In *The Origin of Species*, Darwin emphasized over and over again the structural or functional imperfections of the living world....

If one wanted to play with a comparison, one would have to say that natural selection does not work as an engineer works. It works like a tinkerer—a tinkerer who does not know exactly what he is going to produce but uses whatever he finds around him whether it be pieces of string, fragments of wood, or old cardboards;... with odds and ends. What he ultimately produces is generally related to no special project, and it results from a series of contingent events, of all the opportunities he had to enrich his stock with leftovers. As was discussed by Levi-Strauss, none of the materials at the tinkerer's disposal has a precise and definite function....

From an old bicycle wheel, he makes a roulette; from a broken chair the cabinet of a radio. Similarly evolution makes a wing from a leg or a part of an ear from a piece of jaw.[3]

Neither Jacob nor Vogel was an evolutionary naysayer or fringe figure. Vogel was a world leader in biomechanics and an unwavering evolutionist. Jacob won the Nobel Prize, and his aforementioned "Evolution and Tinkering" is a famous and influential essay in the field of evolutionary biology. Both scientists frankly declare, as did Darwin himself, that the evolutionary process is severely constrained. This means that if such a process did indeed produce all the variety of our biosphere, then the living world should not merely contain cases of dysfunction; it should be marked top to bottom by substandard designs, ones manifestly inferior to those of skilled human engineers.

Lents, Hafer, Dawkins, and many others get it, and they have gone in search of supporting evidence. Lents, for example, has written an entire book, *Human Errors*, detailing what he regards as hundreds of design faults in the human body.[4] Hafer, in her book *The Not-So-Intelligent Designer*, claimed the human eye would deserve an "F-grade in any decent design class."[5] Similarly, Dawkins, in *The Greatest Show on Earth*, has a whole section on "Unintelligent Design" in biology. He asserts that the eye is the "design of a complete idiot," that the "arteries leaving the heart" are "a haphazard mess," and that one branch of the laryngeal nerve takes "an astonishing detour" that "is a disgrace."[6] They aren't simply complaining. Their point is that this is precisely what we should expect from the evolutionary process.

I would add that I totally agree that bad design is what we should expect *if* evolutionary theory were true. The problem for Dawkins, Lents, and Hafer is that their claims of poor design dissolve under close inspection. Indeed, many of the examples of ultimate engineering I give in the pages that follow include the very examples evolutionists regularly give as instances of bad design.

Intelligent Design Predicts Superior Design

In contrast to evolutionary theory, the intelligent design paradigm predicts that biology will not be found to be routinely characterized by bad design. As with human designers, an intelligent designer of the natural world would not face the key constraints faced by the evolutionary process. Most fundamentally, an intelligent designer, unlike evolution, can employ foresight to envision a solution well beyond anything in existence at the time, and then set about making that envisioned solution a reality.

And as this book's title makes clear, I will go further than claiming an absence of bad design. I will argue that biology contains design that is far superior to human technology—design that is, in fact, ultimate engineering. By this I mean design at the limit of what is possible. Take the lubrication system in animal joints. It's not just optimal. It's at the limit of what is physically possible and far better than the best human lubrication system.

This expectation of ultimate engineering fits comfortably with the intelligent design paradigm, and all the more so when reinforced by a specifically theistic design paradigm, for if the whole universe—including its laws and materials—is understood to have been made by an intelligent designer, as theism holds, then it follows that this designer would possess intimate knowledge of how to use those laws and materials to produce designs at the limit of performance.

Of course, organisms can have imperfections due to genetic disorders. However, genetic disorders are the result of the decay that we see in all living organisms, and indeed in all engineered systems in the world, no matter how brilliantly designed. What I am suggesting is that the human body and other organisms were initially made without these genetic flaws, and that the flaws arose over time. Individual organisms also develop imperfections as they age. But again, even the best engineered systems wear out as they age. So when I speak of ultimate engineering, I am describing organisms free of defects from either genetic decay or aging.

Over the last forty years I have studied engineering and biology in the lab and published various research papers on the science of design in engineering and the living world. I also have been editor of a journal of bioinspired design (and guest editor of another) and have read hundreds of research papers that describe brilliant design in nature. So when I say that it is clear to me that the biological realm is brimming with design sophistication not just superior to human technology but near the limit of what is physically possible, I do so from a wealth of observational experience.

We have before us an important test case: Is biology worse than human technology, as evolutionary theory anticipates, or is biology better than human technology, as the design paradigm summarized above anticipates? As we will see, the technology found in living things is consistently better than human technology, thus strongly supporting design.

A Macro-Sized Problem for Evolutionary Theory

To his credit, evolutionist Andrea Rinaldi, quoted in the opening paragraph above, has not fallen into the error of claiming bad design in biology. But in recognizing ultimate engineering in the living world, he faces a challenge. If evolution is a trial-and-error process prone to generate suboptimal designs, how can he, an evolutionist, explain the many cases of advanced engineering in biology? He puts his money on what he sees as the wonder-working powers of natural selection. "From the sophisticated ventilation system of a termite mound to the

tensile strength of spider silk," he writes, "nature has invariably selected the most effective designs through billions of years of relentless
evolutionary pressures."[7]

Here I part company with Rinaldi. Such sophisticated and effective designs, I will show, cannot be explained by "relentless evolutionary pressures." A fundamental weakness of evolution is that natural
selection can create nothing. It can only "select" from among the novel
mutations that happenstance presents it. Some will put their hope in
vast amounts of time to make up for the constraints of evolution, but
that is wishful thinking. The constraints don't just render evolution
slow; they set hard limits on how much change is possible. As we will
see, the actual scientific evidence is showing more and more that only
purposive design can engineer new designs.

To be sure, the idea of small-scale adaptation, or what the great
evolutionary biologist Theodosius Dobzhansky dubbed microevolution, is a well-established fact of science, and there is much good
research in this area. (Interestingly, there is growing evidence that
many types of adaptation are far from random, but instead are due to
various internal pre-programmed adaptive capabilities.[8]) We have no
reason to doubt, for instance, that Darwin's finches, with their range of
different beak shapes, are a classic example of microevolution. Darwin
himself did excellent research on adaptation with many organisms.
The problem is that already in Darwin's day, and increasingly since
then, considerable evidence has pointed to the conclusion that microevolution cannot be reasonably extrapolated to macroevolution.

As Dobzhansky himself conceded, macroevolution has not been
observed.[9] And the more than eighty years of research since he made
this admission have not been kind to the micro-to-macro assumption.
Various findings have only exacerbated the gulf between microevolution
and macroevolution. This gulf will become especially apparent when, in
the coming chapters, we delve into various instances of ultimate design.

Friends in High Places

Those of us who want to follow the evidence without any constraining
ideological commitments to evolutionary theory can take courage from

a steady trickle of brave individuals who have publicly broken with orthodox evolutionary belief—including internationally distinguished scientists such as German paleontologist Günter Bechly, Yale's David Gelernter, and Dean Kenyon, a pioneer in the field of abiogenesis.

Support from distinguished scientists is hard to beat, but friends in other high places are nice, too. In 2000 I published a book, *Hallmarks of Design*, showing that intelligent design is supported by extensive scientific evidence. A copy was given to Prince Charles (now King Charles III), and to my amazement he wrote to tell me he liked the book. He also said he wanted to quote from its foreword (by biologist Alan Linton) in his Reith lecture, which would be broadcast to 90 million people. Permission, as you may have surmised, was readily granted. Here is the relevant passage from the lecture:

> As Professor Alan Linton of Bristol University has written: "Evolution is a manmade theory to explain the origin and continuance of life on this planet without reference to a Creator." It is because of our inability or refusal to accept the existence of a guiding hand that nature has come to be regarded as a system that can be engineered for our own convenience or as a nuisance to be evaded and manipulated, and in which anything that happens can be fixed by technology and human ingenuity.[10]

After the speech, I told one of my university colleagues that I thought Prince Charles was very brave. The colleague replied, "No he's not brave. He cannot be demoted or fired!"

I partly agreed. While Darwin-dissenting scientists have been fired or hounded out of positions after coming out as skeptics of modern evolutionary theory, there was no danger of Prince Charles being sacked as Prince of Wales. However, the prince's criticism of evolutionary theory did cause quite a stir in the national press in the UK, and he was raked over the coals by various scientists who insisted that he, a non-scientist, had no place questioning the assumptions of modern science. He certainly could have made things easier for himself by skipping the whole subject and restricting himself to ribbon cuttings and such. I, for one, was grateful and encouraged that he spoke up, highlighting what any objective investigation of the historical record makes abundantly clear, namely that a central motivation of evolutionary theory, from Darwin on, has been to rule God out of origins.

King Charles is not a scientist, but he recognizes that worldview plays an outsized role in evolutionary science. And like a lot of people, many scientists included, he is convinced that an even-handed examination of the natural world reveals that a divine hand is behind the origin of nature. I was pleased that he seized on that passage from the foreword to *Hallmarks of Design*, because I am convinced that it is a grave mistake to restrict origins science to strictly materialistic causation. True science should be able to follow the evidence to the best explanation, not merely to the best materialistic explanation.

King Charles and Professor Alan Linton are not the only high-ranking people to surprise me with support for intelligent design. I have spoken with several senior biologists who have confided in me that they have serious doubts about macroevolution but would not dare to air them publicly for fear of damaging their careers. I will relay some very revealing conversations with these biologists in the second part of this book.

Biological Marvels of Engineering

If detailed arguments about origins are not your cup of tea, not to worry. The bulk of this book isn't taken up with them. The focus instead is a series of extraordinary engineering feats found in the living world—in the human body especially—feats that anyone can appreciate regardless of worldview commitments. The first five chapters explore the engineering genius evident in the foot, knee, wrist, fingers, and spine. Then we will move up to the head to examine even more sophisticated engineering marvels, including our hearing and vision. We also will look at skin, seemingly simple but in reality a marvel of design sophistication. Additionally, we will consider the biomechanics of birth, of muscles and tendons, and of three physiological systems unmatched by anything in the realm of human technology.

In the final few chapters we will dive deeper into a comparison of evolutionary theory and intelligent design. When we remove ideological constraints and social pressures, and simply follow the evidence, what do we find? Was the origin of these various biological marvels the work of Richard Dawkins's "blind watchmaker," unguided evolution, or was a seeing and designing mind involved?

For me, studying the many examples of ultimate design reviewed in the pages that follow has been fascinating, and a great way to appreciate the awesome wonder of nature. One aim of this book is to convey some of the fun and excitement of biomechanics research. Most of the examples I give are from my research in biomechanics, carried out at Bristol University and Cambridge University. That work has led to my publishing more than 200 scientific papers on the science of design, and the content of most of these has contributed to the arguments in this book. My hope is that even scientists fully committed to the evolutionary paradigm will find much of value in these explorations. My greater hope is that these explorations will encourage such individuals to consider the possibility that an unbending allegiance to evolutionary materialism has exhausted its usefulness as an explanatory tool.

ULTIMATE ENGINEERING
IN HUMAN BIOMECHANICS

1. FOOT AND ANKLE

Two Views of the Human Foot and Ankle

Leonardo da Vinci is often quoted as saying, "The human foot is a masterpiece of engineering and a work of art." There appears to be no verification that the quotation actually came from da Vinci. However, the enduring popularity of the quotation testifies to the fact that many writers have had great respect for the engineering of the human foot. The quotation's popularity likely also benefits from its being very much what one would expect from da Vinci. He was, after all, well known to have deep admiration for the workings of nature and, as a pioneering anatomist, studied the human body in meticulous detail, as shown by his series of masterful drawings.

The more basic question is whether the statement is true. I have studied the biomechanics of the human foot for three decades and can testify that contemporary research has unequivocally confirmed that the human foot is indeed a masterpiece of engineering.

In studying the foot, I was awestruck by the level of sophistication and precision design. In 2023 I published a paper in the journal *Biomimetics* on an advanced prosthetic foot design developed by my research team at Bristol University.[1] The design described there is one of the first to copy the individual ankle bones and ligaments of the human foot. A key finding of my paper was that the human foot is far superior to any prosthetic foot that engineers have produced. I also explained that the best way to improve prosthetic feet is to copy the design features of the human foot. Lest anyone conclude that this thesis was unacceptable to the biomimetics community, I should note that the paper was selected as the "Editor's Choice" for the journal issue.

You can find many examples of scientists giving high praise for the design of the ankle-foot complex. An established biomechanics textbook states that the structures of the foot "work in perfect synchronization."[2] Researchers in bioinspired design report that the foot's ability to rapidly fine-tune stiffness maximizes efficiency of locomotion.[3] In a review of the biomechanics of the ankle joint, researchers describe the ankle as having a high degree of stability and robustness despite very high loading.[4] Researchers in sports biomechanics describe how its unusual combination of stiffness and flexibility yields the "nearly effortless human gait."[5] A medical educational website practically gushes over the clever engineering that affords this combination:

> The midfoot is superbly constructed for ambulation. The tarsal bone articulations have capability of motion that allows for twisting of the foot and pronation/supination. However, during the push-off phase of walking, a stable midfoot column is more important than flexibility. This goal is achieved via the "windlass mechanism".... A windlass is a mechanism commonly used on boats. It is a cylinder, which, when cranked, tightens a rope attached to a sail, anchor, and so on. The plantar fascia acts as a windlass of the foot. It is connected to the calcaneus and the digits. During the push-off phase of ambulation, dorsiflexion of the toes causes the plantar fascia to wrap around the metatarsal heads (i.e., the cylinder), which tightens the fascia, pulling the tarsal bones together and creating a stable column across the midfoot.[6]

Despite universal praise for the design of our ankle-foot complex by experts in the biomechanics community, evolutionist Nathan Lents describes the ankle bones as among "the most obnoxious example of bones for which we have no use."[7] Paleontologist and evolutionist Jeremy DeSilva also sees poor design. The title of his lecture on the topic says it all: "Starting Off on the Wrong Foot: How Our Ape

Ancestry Predisposes Us to Foot and Ankle Maladies."[8] He spoke on the topic at the 2015 AAAS annual meeting. That year's conference theme was "The Scars of Evolution."[9] It would be difficult to more effectively distill the fact of evolutionary theory's expectation of bad design than that conference name.

The perspective could not be more at odds with the facts. So why adopt such a view? Because evolutionary theory predicts poor design. It does so, according to DeSilva, because there was not enough time to evolve the grasping structure of our purported tree-climbing ancestor into the stable lever structure needed by the human foot.[10]

I agree with DeSilva and Lents that evolutionary theory predicts the human foot should be a very poor design. The problem for them is that the scientific evidence points to its being a masterpiece of engineering design. Lents and DeSilva make the mistake of following their evolutionary paradigm rather than the scientific evidence.

The Incredible Multifunctional Foot

The unique human foot gives humans incredible agility in activities such as walking, running, jumping, climbing, and dancing. To achieve such agility, the foot performs an amazing range of functions, which requires the foot to be both stiff and flexible—competing requirements very difficult to achieve in one structure. On the one hand, the foot has to form a very stiff lever for pushing off the ground in running and walking. On the other hand, it has to become very flexible when it lands.

Engineers know it is extremely difficult to quickly adjust and fine-tune stiffness, but this is exactly what the human foot does. It is constantly reconfiguring itself during running and walking to change stiffness.[11] The reconfiguration is carried out by changes in muscle tension and through ligament tightening, especially the plantar fascia ligament, which runs along the bottom of the foot. The feet also have to withstand large deflections when stretching, as when running over rough terrain or playing sports like tennis. The feet also have to perform static strength and balance functions such as standing on one or two legs and supporting the weight of the whole body plus any additional weights being carried.

To pack so much functionality into such a compact design is truly impressive. The foot is also remarkably robust. The feet are able to take tens of millions of steps over an eighty-year lifespan as well as cope with occasional injuries and strains.

All this is why engineers are in awe of the human foot.

Figures 1.1, 1.2, and 1.3 give a glimpse of the immense complexity of the human foot. The foot has twenty-six bones, more than a hundred ligaments, and some twenty-nine muscles, including ten originating outside the foot and around nineteen internal muscles. Other important parts include the skin, cartilage, blood vessels, nerves, and bursa, the latter for lubricating tendon movements. There are also thousands of nerve endings in the feet, which play a crucial role in providing feedback on the forces and movement of the foot. And all these many parts are masterfully integrated into a compact assembly.

The bones of the foot are shown in Figure 1.1. The ankle consists of the hindfoot and the midfoot with a total of seven bones. The forefoot contains nineteen bones. There are around thirty-three precise joints in the foot between individual bones. Each joint is held together tightly by ligaments that are both tough and flexible. Some of the main ligaments of the foot are shown in Figure 1.2.

The ankle joint can move the foot in three different ways, as shown in Figure 1.3. The main movement is up and down. The joint can also pivot left and right, and it can tilt inwards and outwards. Pronation is a rotation that combines all three.

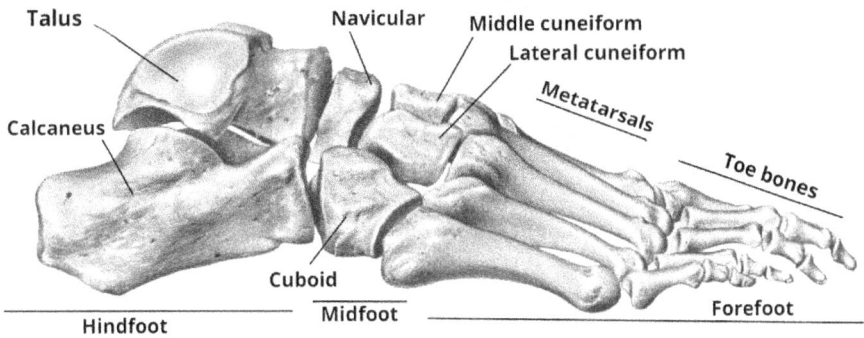

Figure 1.1. Bones of the foot.

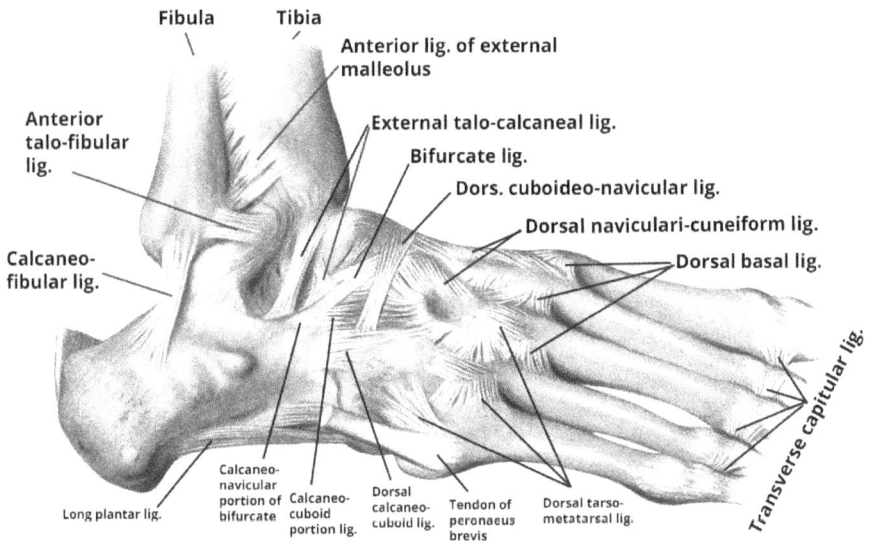

Figure 1.2. Some of the main ligaments of the foot and ankle.

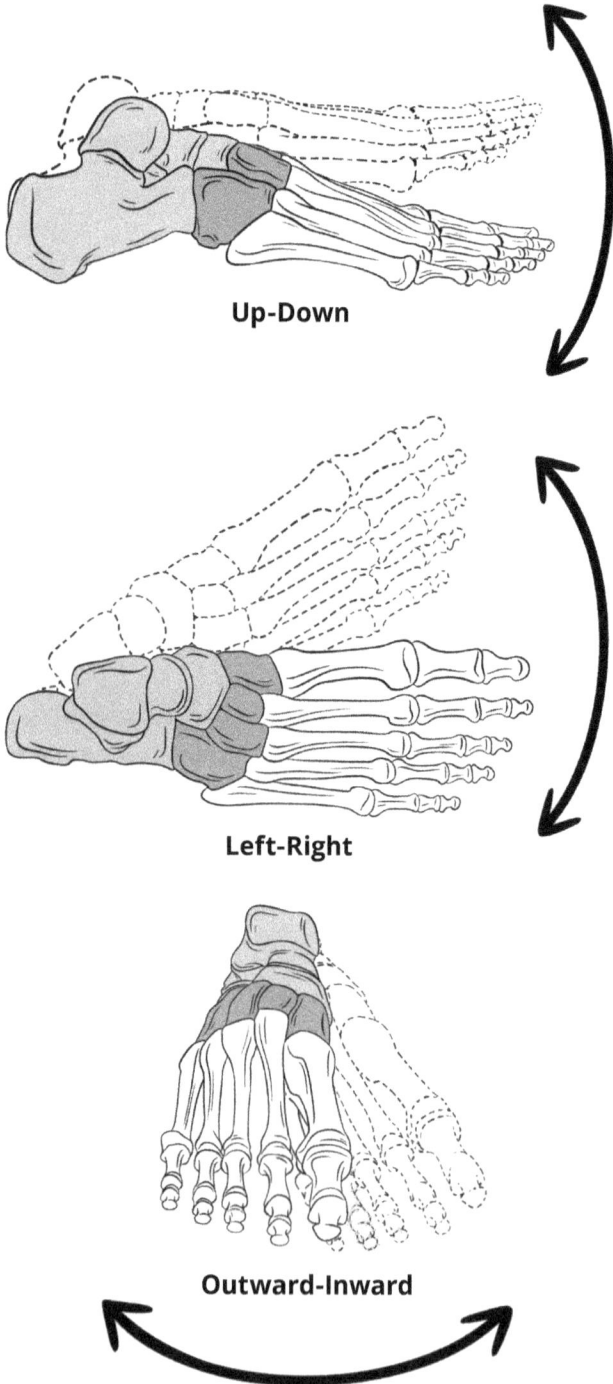

Up-Down

Left-Right

Outward-Inward

Figure 1.3. Three ankle joint movements.

The human foot is unique in design and easily differentiated from an ape's foot. The ape foot is like a flat hand for grasping, with the big toe pointing outwards, like a thumb. The human foot has a completely different arched structure with a stiff big toe pointing forwards.

A Triple-Arched Structure: A Design Masterpiece

The arch-structure of the human foot is worth pausing over—an ingenious triple-arched structure that has astounded engineers with its precision and performance across a range of crucial functions.

The features and functions of these arches are summarized in Figure 1.4. The medial arch extends from the heel (calcaneus bone) to the three biggest toes and includes all the bones of the ankle and forefoot that line up with those toes. The lateral arch extends from the heel to the two smallest toes and includes all the bones of the ankle and forefoot that line up with those toes. The transverse arch connects the other two arches and is formed by some of the small bones in the ankle as well as some of the bones in the forefoot.

The three interconnecting flexible arches give ideal three-point contact with the ground. All three arches are able to deform and flatten to absorb shock as well as store and release energy. There are also specific functions for each arch. The medial arch is the strongest arch and can form a very stiff lever for push-off in walking and running. It has two contact points with the ground, one at the heel and one at the ball of the foot at the base of the big toe. The medial arch has many ligaments that store significant elastic energy during each running stride. The spring ligament is one of the most important of these energy-storing ligaments, hence the name.

The lateral arch creates the third contact point with the ground, at the ball of the little toe, thus maximizing the distance between the two front points of contact, and thus maximizing stability during activities like running. The lateral arch gives stability to the foot, such as when standing on the toes. The transverse arch helps transmit loads from the lateral arch to the medial arch during pronation when the foot rolls from the outside of the foot to the inside.

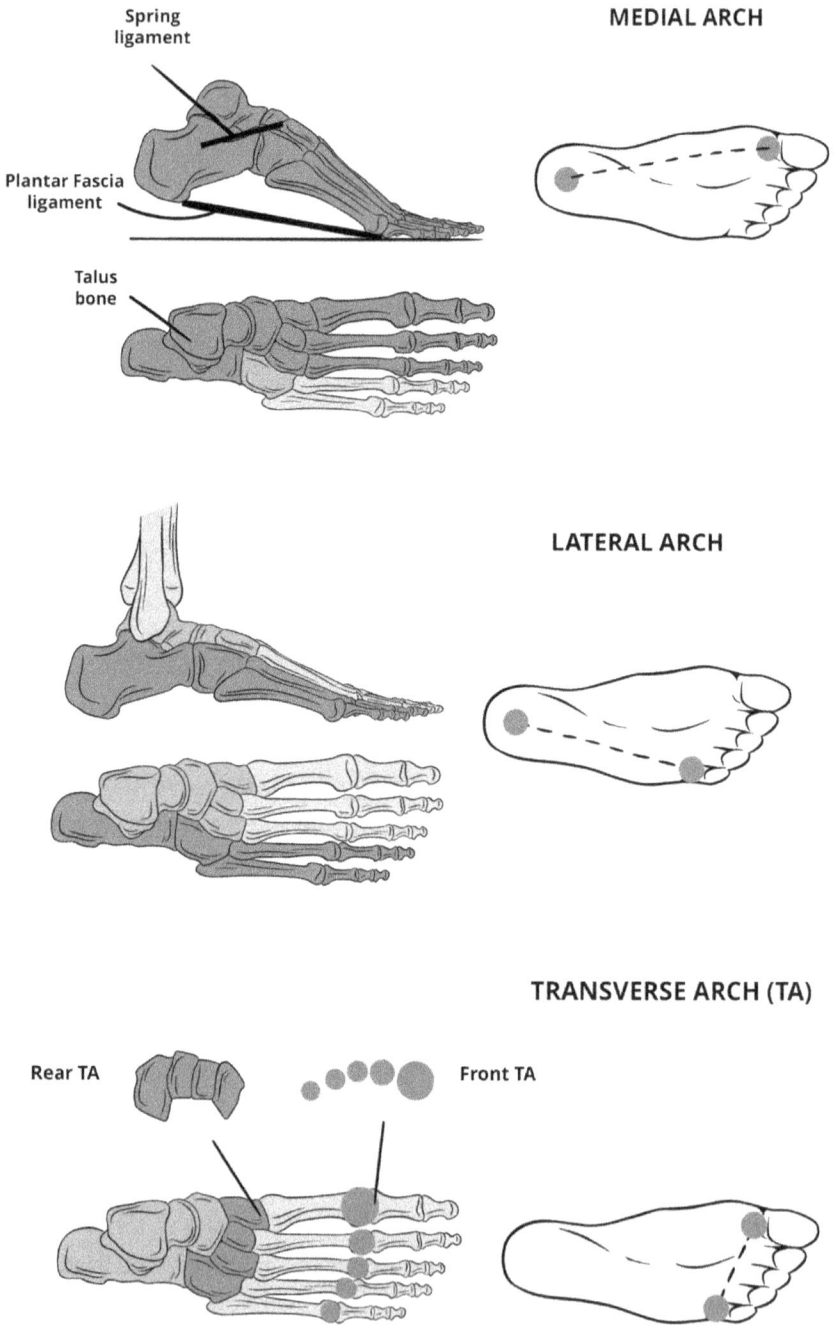

Spring ligament

MEDIAL ARCH

Plantar Fascia ligament

Talus bone

LATERAL ARCH

TRANSVERSE ARCH (TA)

Rear TA

Front TA

Figure 1.4. The three arches of the foot.

Even though the transverse arch is itself flexible for shock absorption, it nevertheless increases the stiffness of the medial arch during push-off. Recent studies have estimated that the transverse arch contributes over 40 percent of the stiffness of the medial arch during push-off.[12] This is an example of how the arches of the feet are multifunctioning. The arch's stiffening effect can be illustrated with a sheet of cardboard. When the cardboard is flat it is easy to bend. When it is curved in the transverse direction to the direction of bending, the cardboard becomes much stiffer.

Biomechanics of Pronation

Pronation is where, when a person is walking or running, the foot rolls from the outside toward the inside after the foot strikes the ground. Pronation gets a bad rap because overpronation in running contributes to inefficiency and injury susceptibility, and some people conflate overpronation with ordinary pronation. But it's only excessive pronation that is a problem. Proper pronation is important because it acts as a shock absorber by creating a soft landing of the foot on the ground.

Because the foot takes the weight of the whole body, the joints in the foot must be very strong. The best design solution for creating a strong and stable joint for pronation is to have an external stabilizing mechanism. This is exactly what the foot has with the fibula bone of the lower leg (a smaller "twin" of the tibia bone) forming a linkage mechanism with the tibia that stabilizes and strengthens the foot in pronation.

In engineering systems, linkage mechanisms are a clever trick for optimizing design. But the foot got there long before any engineer came up with the idea. The foot/lower-leg linkage system is like a double-wishbone car suspension system. The long bars in Figure 1.5, the tibia and fibula, are connected by ligaments such that the lower leg can be modeled by a four-bar linkage with spherical and sliding joints.[13] The result is, among other capacities, a fibula that allows fine-tuning of the moment arm (for mechanical advantage) for the muscles acting on the ankle-foot complex. A second advantage is that the fibula increases the attachment area for muscles, allowing more muscle to act on the joint.

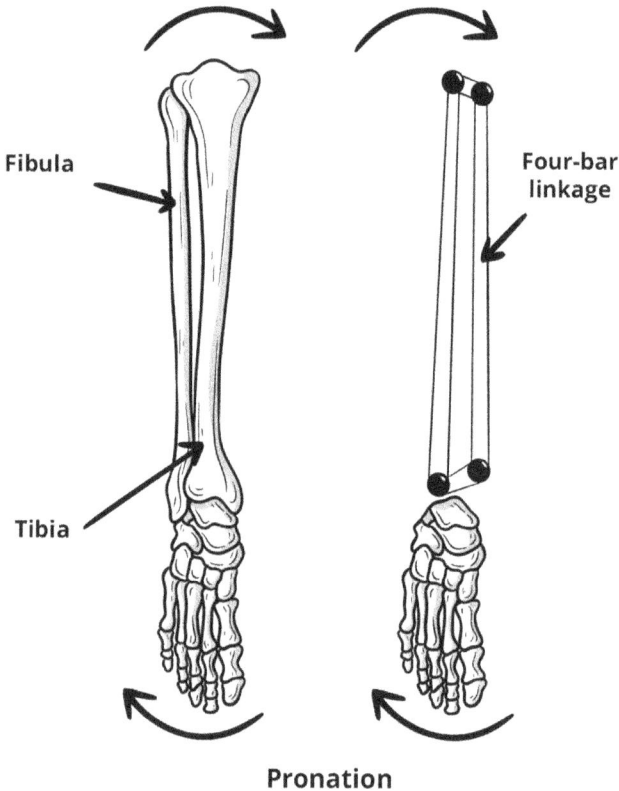

Figure 1.5. The linkage mechanism of the fibula and tibia.

Biomechanics of Standing

The triple-arched structure is a perfect design for standing on one or two legs. Figure 1.6 shows schematics of three different modes of standing: two-legged standing, one-legged standing, and standing on the balls/toes of one foot during the push-off phase in walking or running. Bipedal standing is easy for humans because we can feel the back (heel) and front (ball) of the feet and then put the center of gravity of the body between these two points. In contrast, apes find it difficult to balance on two legs because they do not have arched feet with clear ground contact points. Instead, their feet form a flexible contact patch with the ground that is difficult to control in bipedal standing.

For humans, balance in one-legged standing is achieved by putting the center of gravity of the body inside the triangular footprint formed by the three main contact points of the foot. A separate triangle is formed when one rolls onto the balls/toes of the foot. The end of the big toe forms one of the reaction points, with the other two being the ball of the big toe and ball of the little toe.

It is so easy to take for granted the precision biomechanics involved in standing, walking, and running. In reality, our feet are marvels of multifunctional design.

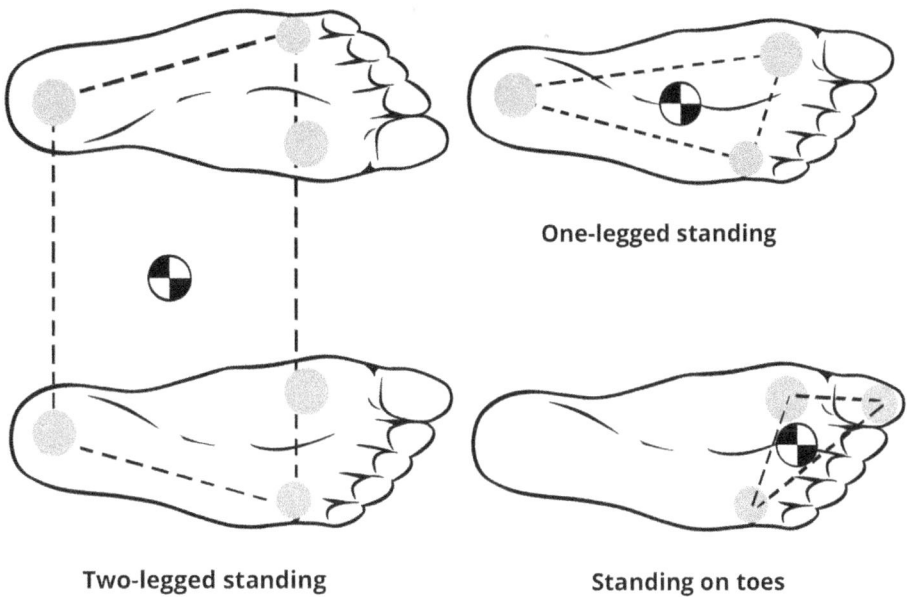

One-legged standing

Two-legged standing

Standing on toes

Figure 1.6. Three standing modes. The ◑ symbol pinpoints the body's center of gravity, which must lie within the dotted boundary lines for stable standing.

Biomechanics of Load-Transfer

During activities such as running, loads must be transferred between the ankle and the toes. The four-bone layout of the midfoot transverse arch (three cuneiforms and cuboid) is optimal for transferring horizontal loading because the bones are in line with the five toes and five metatarsals as shown in Figure 1.7. This creates clear compression-load paths from the toes in the forefoot to the bones in the hindfoot and hence minimizes bending loads.

Bending loads are structurally inefficient because they magnify forces. For example, if you hold a pencil at both ends, it is quite easy to snap the pencil by bending it. However, the pencil is very strong in tension and compression because there is no magnification of forces. The fact that the cuboid bone receives loads from the two smallest toes does not create much bending in the cuboid because the loads are significantly lower in these two toes.

An ingenious feature of the foot that enhances load transfer involves the position of the talus and navicular bones (Figure 1.7). The talus precisely lines up with the navicular so that during push-off the power of the leg muscles is directed to the medial arch. In contrast, when landing on the lateral arch, the forces are not transferred directly through the talus and into the leg, thus reducing shock loads.[14]

Not only does the navicular line up precisely with the three cuneiforms but it also performs a bearing function with the talus bone during pronation movements. (The bearing function involves accommodating relative sliding between the bones.) This is ideal because it creates a one-to-one bone joint, which is much better than having three cuneiform bones sliding against the talus bone. This means that the navicular solves a load-bearing function at the same time as solving a bearing function.[15] Here we see an excellent example of a multifunctioning bone.

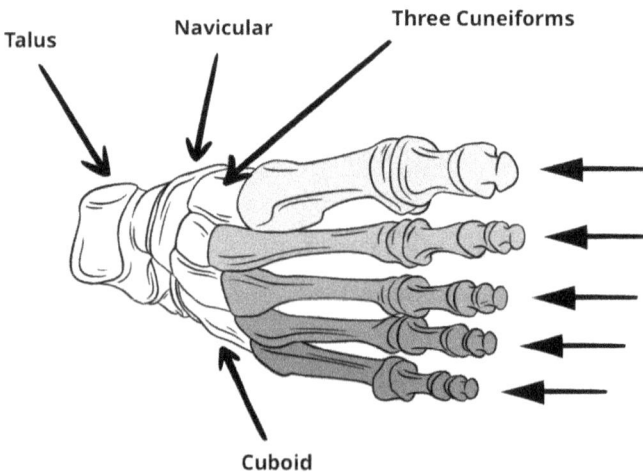

Figure 1.7. Load paths through the foot during running.

Extreme Fine-Tuning in the Foot

The human foot contains incredible levels of fine-tuning and multi-functioning. The bones of the feet not only line up precisely to form load paths and joints as shown in Figure 1.7, but form perfect integrated arches at the same time.

While researching its design, I was many times astonished at the extent of fine-tuning and multifunctioning. Having worked on spacecraft design, I am familiar with the most advanced engineering systems produced in modern times, and yet the human foot surpasses them all.

In the above material we have only briefly considered the finely tuned biomechanics of the foot. There is much more that could be said. And if we were to look at the role of the brain and senses for controlling the foot during running, it would only add to the wonder of the design. If we were to look at the blood flow, muscle actuation, and lubrication of joints, that would add yet more wonder.

A Bioinspired Foot

Figure 1.8 shows the prosthetic foot developed by my research team at Bristol University. One reason we were successful was that we shamelessly copied from the anatomy of the human foot. For example, we used three-dimensional scans of human bones and meticulously reproduced the bone shapes using computer-aided machining and 3D printing technology. These human-like bones were then used in our prosthetic foot.

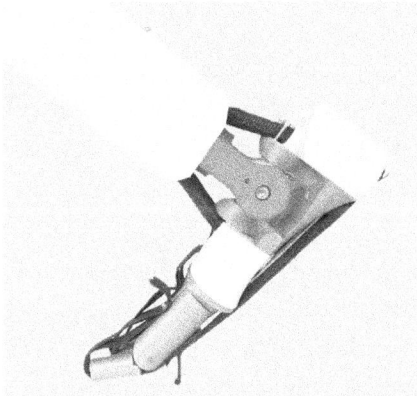

Figure 1.8. A bioinspired prosthetic foot developed at Bristol University.[16]

A particularly fun and fascinating experiment we performed in our research was to make a footprint with our prosthetic foot in a soft pliable material. We hoped that the prosthetic footprint would replicate the three-point contact of the human foot. We also hoped to see imprints of individual toes. Our prosthetic indeed produced three-point contact as well as toe imprints. It's one of the first to reproduce this type of ground contact.

Irreducible Complexity of the Foot

There is a running debate about which culture in antiquity invented and perfected the block-style arch that came to be known as the Roman arch. But none of those cultures can claim the olive wreath, for the designer of the human foot got there long before.

Figure 1.9 compares a Roman arch to the arch of the human foot. The staples and tie in the Roman arch are equivalent to the ligaments in the human foot. Both arches have blocks and a keystone at the top. However, the human foot is more sophisticated than a typical human-made arch because the foot has three integrated arches and three keystones. In addition, the human foot is a dynamic structure that deforms to absorb shock and store energy.

The triple-arched structure of the human foot, an engineering marvel, is unique among mammals and a serious challenge to evolutionary theory. In the fossil record there are no transitional forms between the arched feet of humans and the flat feet of apes. This should not be surprising, because the arches of the foot are irreducibly complex structures that could not evolve step by tiny step. This is because the evolving foot would have to move through a series of intermediates far less functional than either ape feet or human feet, much less functional intermediates that natural selection would readily weed out.

A block-style Roman arch consists of separate wedge-shaped blocks that, under gravity, lock the arch together. A Roman arch is a very efficient form of arched structure because loads are transferred by compression forces rather than bending forces. The design is especially well suited for materials like stone that are stronger in compression than tension. Bone is also stronger in compression than tension, another reason a block-style arch is a good engineering solution for the feet.

In engineering, it is well known that a Roman arch is irreducibly complex, because the arch only functions when the assembly is complete. Notice in Figure 1.10 how a wooden jig is used to hold up the arch during construction. The jig must be in place during the entire construction because the arch cannot support itself until it is complete. In the same way, all the bones of the feet must be in place to function as arches. You cannot have a quarter of an arched foot or half an arched foot. Therefore, the arches of the feet could never evolve in a gradual, step-by-step process.

In Figure 1.9 notice the keystone at the top of the Roman arch. This stone locks the whole arch. The human foot also has a keystone for each of the three arches. The talus bone is the keystone for the medial arch. The cuboid bone is the keystone for the lateral arch. And the middle cuneiform bone of the midfoot is the keystone for the transverse arch. These keystones can be seen in Figure 1.1 above.

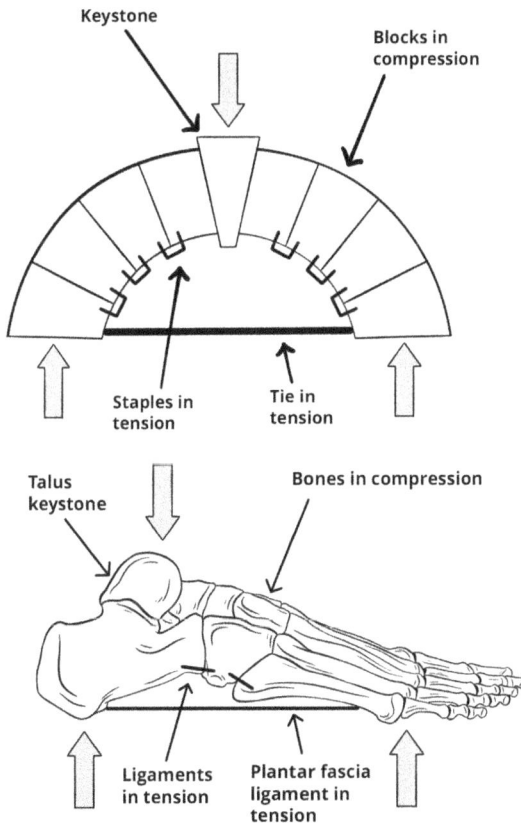

Figure 1.9. Roman block arch (top) and foot arch (bottom).

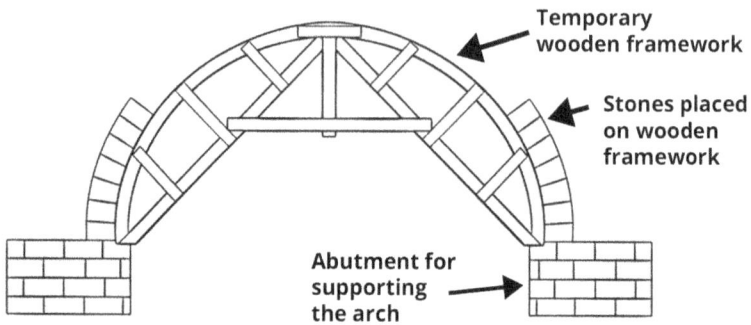

Figure 1.10. Construction jig for a Roman block arch.

One might object that sandstone arches have evolved through the gradual removal of material from the center, eventually leaving behind solid arched structures (as at Arches National Park in Utah), structures that clearly came about through purely blind natural processes. Yes, but those are not strong Roman arches, each made up of several separate stones brought together to form a stable arch. Although striking in appearance, these natural arches are much weaker because they transfer loads through bending rather than compression. The human foot instead employs an advanced version of the Roman arch, one that, like the Roman arches made by human engineers, is irreducibly complex. Therefore, the solid sandstone arches are irrelevant to the engineering challenge faced by an evolutionary process "attempting" to build a simultaneously stiff and flexible foot arch.

The arches of the foot, it should be emphasized, are not the only irreducibly complex structures in the lower body. The linkage mechanism of the fibula and tibia bones, for instance, is also irreducibly complex, because the linkage system must be complete to have a useful function.

Absence of Fossil Evidence for Foot Evolution

A famous fossil claimed as a transitional between an ape-like creature and humans is known as Lucy, identified as the species *Australopithecus afarensis*. Many museums and science books provide recreations of the once-living Lucy standing upright with human-like arched feet.

However, while the fossil remains of Lucy include bone fragments from the arms, vertebrae, pelvis, leg, and skull, there are no bones from the hands and feet. So the anatomy of Lucy's feet is speculation.[17] In the case of other *Australopithecus afarensis* fossils, there is no good evidence for arched feet.

Some have asserted fossil evidence for ape-to-human foot evolution, but such claims have not held up to scrutiny. For example, evolutionist Carol Ward and her fellow researchers claim that a single foot bone (fourth metatarsal) of an *Australopithecus afarensis* fossil resembles the fourth metatarsal of a human biped. "A complete fourth metatarsal of *A. afarensis* was recently discovered at Hadar, Ethiopia," they write, and add that "it exhibits torsion of the head relative to the base…. These features… support the hypothesis that this species was a committed terrestrial biped."[18] But the human foot, recall, contains twenty-six bones, so making such a claim based on the shape of one bone is a very dubious argument. It is like finding a single stone and confidently predicting that it used to be part of a Roman arch. Such a prediction would be totally unjustified because the stone could have been used in many different non-arch applications, such as in a circular stone border around a tree.

The tenuous nature of the evidence did not prevent the popular science media from rushing in to celebrate. A *Science Daily* story announced, "Researchers have found proof that arches existed in a predecessor to the human species that lived more than 3 million years ago." Another in the same online magazine the next day informed readers, "A fossilized foot bone recovered from Hadar, Ethiopia, shows that by 3.2 million years ago human ancestors walked bipedally with a modern human-like foot."[19]

But fellow pro-evolution paleontologists who took the time to test and examine the argument advanced by Ward and her co-authors rejected their conclusion. In 2012, in the mainstream and peer-reviewed *HOMO: Journal of Comparative Human Biology*, P. J. Mitchell, E. E. Sarmiento, and D. J. Meldrum showed how the claims lacked credibility. "None of the correlations Ward et al. make to localized foot function were supported by this analysis," they concluded. "The Hadar 4th MT [metatarsal] characters are common to catarrhines [old world monkeys] that have a midtarsal break and lack fixed transverse or longitudinal arches…. This study highlights evolutionary misconceptions underlying the practice of using localized anatomy and/or a single bony element to reconstruct overall locomotor behaviors."[20]

Interestingly, the popular media did not report on this refutation of Ward's evidence for Lucy walking upright, so the public are still reading the popular articles claiming there is proof that Lucy walked upright, unaware that Ward's evidence has been discredited.

Answering Claims of Bad Design

Experts in biomechanics view the human ankle-foot complex as a masterpiece of engineering. And yet there are evolutionists who insist it is a bad design, the substandard engineering product of that blind tinkerer, Darwinian evolution, only semi-successfully converting ape feet to the bipedal feet of *Homo sapiens*. Evolutionist Nathan Lents well represents this perspective. "Because many of the bones of the ankle do not move relative to one another," he writes, "they would function better as a single, fused structure, their ligaments replaced with solid bone."[21]

It is hard to express the magnitude of the error of that statement.

It is very well known in the fields of medicine and biomechanics that ankle fusions lead to a significant degradation of ankle performance. One hospital report states, "Walking on rough ground is more difficult after an ankle fusion because the foot is stiffer. It is rare to be able to play vigorous sports such as squash or football after an ankle fusion."[22] Another medical report warns, "Because the fused ankle no longer moves, your body may shift more stress to the subtalar joint (below the ankle) or other nearby joints in the foot and knee. Over many years, this can sometimes lead to secondary arthritis in adjacent joints."[23]

There are some pathological conditions, such as tarsal coalitions, which fuse bones in the midfoot, and in such cases foot performance is always significantly degraded.[24]

Lents claims another bad design for the lower leg, insisting that "there is no real reason to have paired bones" in the lower leg.[25] Presumably he's envisioning a revised design in which the fibula bone is dispensed with, and the remaining larger bone (the tibia) becomes still bigger, allowing it to single-handedly manage what previously required the two bones. But Lents's one-bone strategy represents an engineering step backwards. The fibula is well known to be crucial in providing stability to the ankle joint. As one group of researchers explain in the journal *Foot and Ankle International*, "The whole fibula including the head is essential for the stability of the ankle joint complex, and the distal fibula is responsible for stabilizing the ankle mortise during external rotation and inversion."[26]

Another research team explains the importance of the fibula bone: "In recent years, there has been an increasing recognition of the importance of the fibula and the tibiofibular ligaments to the biomechanics of the lower limb as a whole and to the ankle joint in particular."[27]

Lents's errors regarding the design of the ankle and lower leg show the dangers of following an evolutionary paradigm rather than well-established scientific knowledge.

The Actual Reasons for Foot Injuries

According to Lents, "There is a reason that twisted and sprained ankles are so common: the skeletal design of the ankle is a hodge-podge of parts that can do nothing except malfunction."[28]

Here Lents makes the error of ignoring another possible culprit: misuse rather than bad design. Research by sports scientists reveals that when human foot and leg joints suffer injury, the culprit is generally misuse.[29] Common types of misuse include ramping up training loads too quickly and carrying out high impact sports when overweight.

Another reason joints go bad is that today people regularly live well into their eighties and nineties. We also live at a time where people have sedentary, unhealthy lifestyles. If you look at past generations, when people were much more active and life expectancy was lower, joint problems were much less common.

Lents's claim that the "ankle can do nothing but malfunction" is worse than an exaggeration. What is noteworthy is not the frequency of ankle problems but the frequency of the ankle working. In the case of healthy adults who are not overweight, it is remarkable how robust and long-lasting the ankle joint can be.

I can give a personal testimony on the matter. I am a keen long-distance runner and run virtually every day. During my adult lifetime I have run approximately 80,000 miles, including running sub-three-hour marathons during my sixties, and I have had very few problems with my ankles. Granted, my body type and lifestyle are well-suited for logging so many miles decade after decade. But how many of us know individuals who are clearly not cut out to be competitive distance runners who nevertheless have ankles that effectively support them through many decades and tens of thousands of miles of bipedal movement over all manner of surfaces? That ankle joints cope with so much use, and such varied use, represents performance far beyond anything engineers can achieve with a prosthetic foot/ankle complex.

Even the best engineered systems wear out over time. If a car is misused or neglected, we can expect it to malfunction, even if it is brilliantly engineered. If a new BMW car is driven into a light post, the damage is not due to poor design, but usually to the reckless actions of the driver. Similarly, if a person neglects to exercise and then one weekend injures himself lifting excessive weight, this is an example of misuse, not poor design.

The same goes for a car that is well-treated but very old. A BMW is an example of excellent engineering, but it will get old and rusty, with various parts eventually wearing out under normal use. In the same way, the human body, although well engineered, will get old and wear out.

Lents sums up his disdain for the foot joint when he states, "No engineer would design a joint with so many separate parts."[30] If he had consulted with engineers who have studied the human foot, he would have learned that engineers are in awe of the design of the foot joint and would love to copy it if they had the materials and technology to make that possible. The design of complex systems requires decades of training and practice, and yet Lents makes profound engineering judgments with no experience of engineering and apparently without consulting engineers.

Lents's claim is like looking at the bevy of gears and other mechanisms in a Rolex watch and declaring that the contraption obviously has too many parts. The casual viewer may be baffled by all those parts, but there are good reasons for all of them. In the same way, there are good reasons for the many parts of the foot and ankle.

No Better Solution Offered

The way to prove that the foot is a bad design would be to present a better design. This the evolutionists have failed to do. When Lents claims that the ankle is a terrible design, he cannot offer a better solution. (His misguided fewer-bones idea is certainly not a better solution.) That is not surprising, because no engineer on the planet knows of a better solution.

Darwinists, it seems, rush in where engineers prudently fear to tread. Lents's advice is *fewer bones*. Another Darwinist, also unencumbered by the expertise of engineers, goes further. Alice Roberts suggests that humans would be better off having feet like an emu,[31] a large flightless bird that can run fast and efficiently on two long slender legs, each of which has three clawed toes.

But the legs and clawed toes of the emu serve a fairly limited range of functions (primarily, running and self-defense), whereas human feet are designed for numerous functions.

To perform the running function, the emu requires a relatively simple foot without an arched structure, without much flexibility, without a large ground footprint, and without much ability to move the ankle joint in pronation. Therefore, having emu feet would result in many problems for humans. Standing, balancing, and reaching would be harder because the emu foot has a smaller footprint and fewer directions of movement. Also, movements like squatting would be very difficult. Playing sports would be much harder because it would not be possible to stand on the front of the foot and change direction with precision. Ball sports that involve running would be virtually impossible, as would many of the real-world survival activities of our technologically less-advanced ancestors, for which many of today's sports function as a rough proxy.

What Roberts has failed to understand is that humans have all-round agility in the way that they can reach, bend down, stretch, climb, crawl, and run. This all-round agility requires a multifunctioning foot, not an emu foot designed for a much more limited range of functions.

Saying that humans should have feet like emus is like saying a Swiss army knife would be better off having just a single knife and bottle opener. It would be lighter, more compact, more affordable. But of course, such an argument ignores the fact that the great attraction of the Swiss army knife is its multifunctionality.

Devolution from an Optimal State

Both evolutionary theory and the theory of intelligent design (ID) recognize that nature is in a continual process of change. Both theories also offer explanations for both function and dysfunction in biology. But then the two theories part ways.

Evolutionary theory tends to view examples of dysfunction as cases of the blind trial-and-error process of evolution doing the best it can, often stuck with a poor hand of cards. The supposed dysfunction is seen as a result of forward-developing evolution that can't go back to square one but must re-engineer some existing design fitted for a quite different purpose.

In contrast, design theorists don't anticipate poor design in biology, are open to evidence of ultimate engineering, and understand that even the best systems devolve or degrade over time, or suffer from misuse. After all, machines and other complex systems, even the best, degrade over time, so it's what we should expect.

Thus, the fact that organisms have accumulated genetic disorders (devolution) over successive generations does not mean that the organisms were badly designed in the first place. When judging the quality of some work of engineering, what engineers look for isn't the unrealistic absence of all degradation over time but how well the system resists degradation. The ability of organisms to minimize degradation over numerous generations is remarkable, and how this is achieved is even more so. Investigators, for instance, have now discovered various eerily clever nanobots at the cellular level dedicated to error prevention and correction.[32]

What about diseases? The diseases themselves may be the product of degradation of microbial forms in the biosphere, or of immune systems previously robust enough to handle a disease but now vulnerable to it due to devolution. In any case, diseases are a form of attack, and even the most skillfully engineered human systems are not invulnerable to debilitating attacks of one kind or another. If you are driving a BMW and a tree falls on the car, or bad gasoline fouls the car's engine, the problem is not the car design but a hazard. One might insist that a properly wise designer would have removed all threats from living organisms, and from humans in particular; but that is a theological claim rather than one rooted in either science or the common experience of engineers and other designers. Since it is a common theological claim leveled by many evolutionists, however, we will consider it later in the book.

An Extraordinary Feat

Science and engineering together have made clear that the foot is a masterpiece of engineering. What I find extraordinary about its design is that the same set of bones work so well for the very different functions of strength, stiffness, and flexibility. That is outrageously good design.

The claims of bad design by Lents, Roberts, DeSilva, and like-minded evolutionists reflect a serious lack of knowledge of engineering principles and the biomechanics of how the foot works. Their misguided claims of bad design illustrate the debilitating effects of imposing an evolutionary paradigm on the ankle-foot complex rather than following the scientific evidence.

There is a need for students to be taught what experts in biomechanics have discovered about the foot and ankle (outstanding design) and not just what evolutionary philosophy says about it (bad design). It would also be good if students learned how intelligent design theorists, contra Darwinian expectations, anticipated precisely what has been discovered regarding the ankle-foot complex: evidence of advanced, precision engineering.

2. THE KNEE JOINT

Two Views of the Knee Joint

"It is remarkable," comments S. D. Masouros in the journal *Ortho-paedics and Trauma*, that knee joints "undergo millions of cycles of loading without showing symptoms of wear."[1] Masouros's finding, based on research at the world-leading biomechanics labs at Imperial College in London, underscores how a healthy knee, if not abused, can be used for decades without significant wear. No engineered knee prosthetic approaches that level of performance despite the use of the best available materials and the best human technology.[2] All prosthetics show signs of wear after just a few years of use.

Like my colleagues in biomechanics research, I am in awe of the brilliant design of the knee joint. I have been researching it for thirty years, including testing human cadavers and designing prosthetic and robotic joints. I have published several scientific papers on prosthetic and robotic knee joints, including two in a journal of the American Society of Mechanical Engineers, the leading publisher of mechanical design in the world.[3] I have shown in my papers how the human knee is a brilliant design, with an optimal combination of two key performance measures—high range of movement and high mechanical advantage (high moment arm for the muscles).

One feature of the knee joint that engineers have struggled to copy is its exceptional range of motion (ROM). Medical consultant Simon Garrett describes the knee joint as "your body's engineering marvel. This design masterpiece allows for a remarkable range of motion—from walking and jumping to those envy-worthy dance moves."[4]

Despite the high praise given to the knee joint by experts in the field, Nathan Lents insists that it's a bad design. "The problem is due to incomplete adaptation. Nowhere is this clearer than in the human knee," he asserts. A bit later he adds, "The anatomical adaptation to upright walking never quite finished in humans. We have several defects that are the result of the failure to complete the process.... The ACL is vulnerable to tearing in humans because our upright bipedal posture forces it to endure much more strain than it is designed to."[5]

Lents makes these comments not because of any clear and compelling scientific evidence but because evolutionary theory predicts poor design. I fully agree with him that the evolutionary paradigm predicts that the knee should be a very poor design. The problem for Lents is that the scientific research points to the knee being a brilliant design and far beyond anything engineers have produced. Compared to the human knee, the best prosthetic knees have a smaller range of movement as well as a much shorter wear life.

Like the foot, the knee joint is an incredible multifunctioning apparatus. The knee has two main joint functions—large flexion-extension movement and some axial rotation between the femur and tibia. The large range of movement in flexion-extension is necessary for activities like running and jumping. A small amount of knee axial rotation is important for activities like skiing and changing direction in running.

In addition to its two joint functions, the knee has two important structural functions: high strength and locking in the standing position. The knees must be strong because when a person is running and jumping, each knee can experience forces over six times the weight of the body. The knee also needs to be able to absorb shock loads and lock in the standing position. All this, and the knee is able to undergo millions of movements over an eighty-year lifetime.

An Ingenious "Floating Joint"

The knee is a "floating joint" because, unlike most other joints (e.g., the hip joint), the knee has no fixed center of rotation but is free to rotate and roll. The floating nature of the joint—rarely seen in human technology—is an ingenious design, giving the knee joint an exceptional range of movement in flexion and extension.

Figure 2.1 summarizes the special anatomy of the knee joint. The front view shows the patella (kneecap) and the quadriceps muscles. The side view shows the femur, tibia, and patella bones. The femur possesses two separate condyles (rounded protuberances) with a gap in between to give space for the two cruciate ligaments that join the bones together and guide their motion. Additionally, there is a sliding joint located between the patella and a groove in the femur.[6]

The reason the floating joint is so clever is that it allows the joint to excel in two ways that normally are not possible simultaneously: (1) large load capacity and (2) large range of motion. They're not normally possible together because to carry high loads, the bones need large curves at the joint to avoid high stress, and these restrict the range of motion to a small angle. The floating-joint design brilliantly overcomes this problem, allowing the femur to roll over the

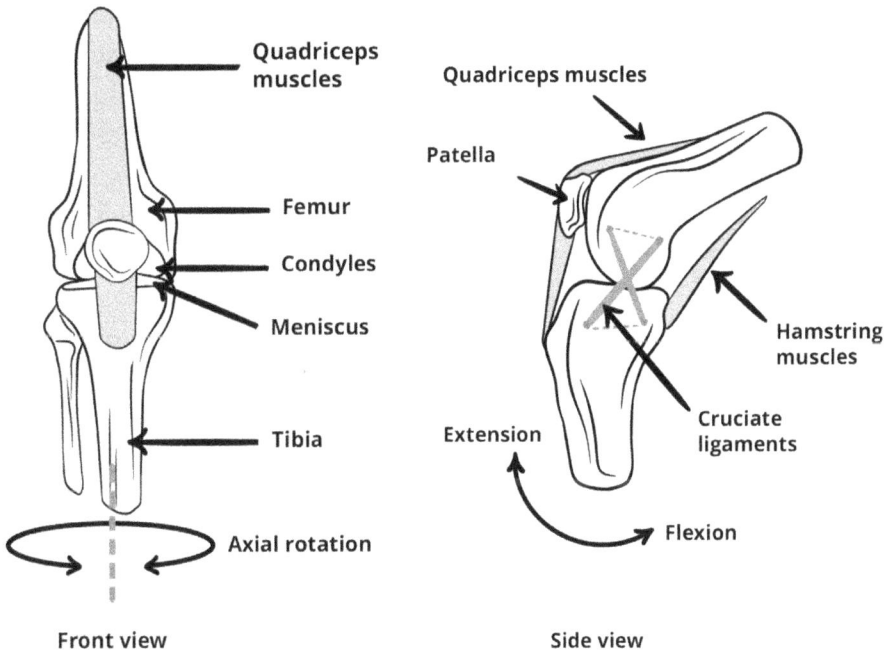

Figure 2.1. Anatomy of the knee joint, front and side views.

tibia with a moving center of rotation, thus affording a great range of motion.

This clever solution requires some seriously advanced engineering. It requires (1) a complex, fine-tuned geometry in the bones; (2) a sophisticated four-bar linkage system; and (3) a special meniscus structure between the femur and tibia.

Biomechanics of the Four-Bar Linkage Mechanism

The bone geometry and motion of the four-bar linkage are shown in Figure 2.2. The cross links are formed by the anterior and posterior cruciate ligaments (ACL and PCL). The parallel bars are effectively formed by the femur and tibia. (These bars are represented schematically by dotted lines in Figure 2.2.) In technical terms, the mechanism is an inverted four-bar parallelogram mechanism that forms a rolling hinge.

The four-bar linkage mechanism gives the knee a moving center of rotation that enables the femur bone to roll around the tibia and hence produce large flexion angles. The action of the four-bar mechanism also helpfully increases the moment arm of the quads (mechanical advantage) during squatting. This means that the lower you squat, the more force you get from the quads—which is exactly what you need when squatting low.

When engineers first studied the knee joint, they were astonished to find such a creative solution to a complex problem. No engineer had ever thought of such a system.

The diagrams in Figure 2.2 grew out of work (during a three-year period) by my PhD student A. C. Etoundi and me studying the knee, after which we were all the more amazed at the precision of the design. The research also helped us appreciate the function of the patella. It forms a low-friction joint that reacts to the loads generated from the quadriceps muscle-tendon group. It also increases the quadriceps' mechanical advantage.

One feature of the knee joint that astonished us is that the curved profile of the femur and tibia has to exactly match the motion of the four-bar mechanism; otherwise the knee will lock up. And when I

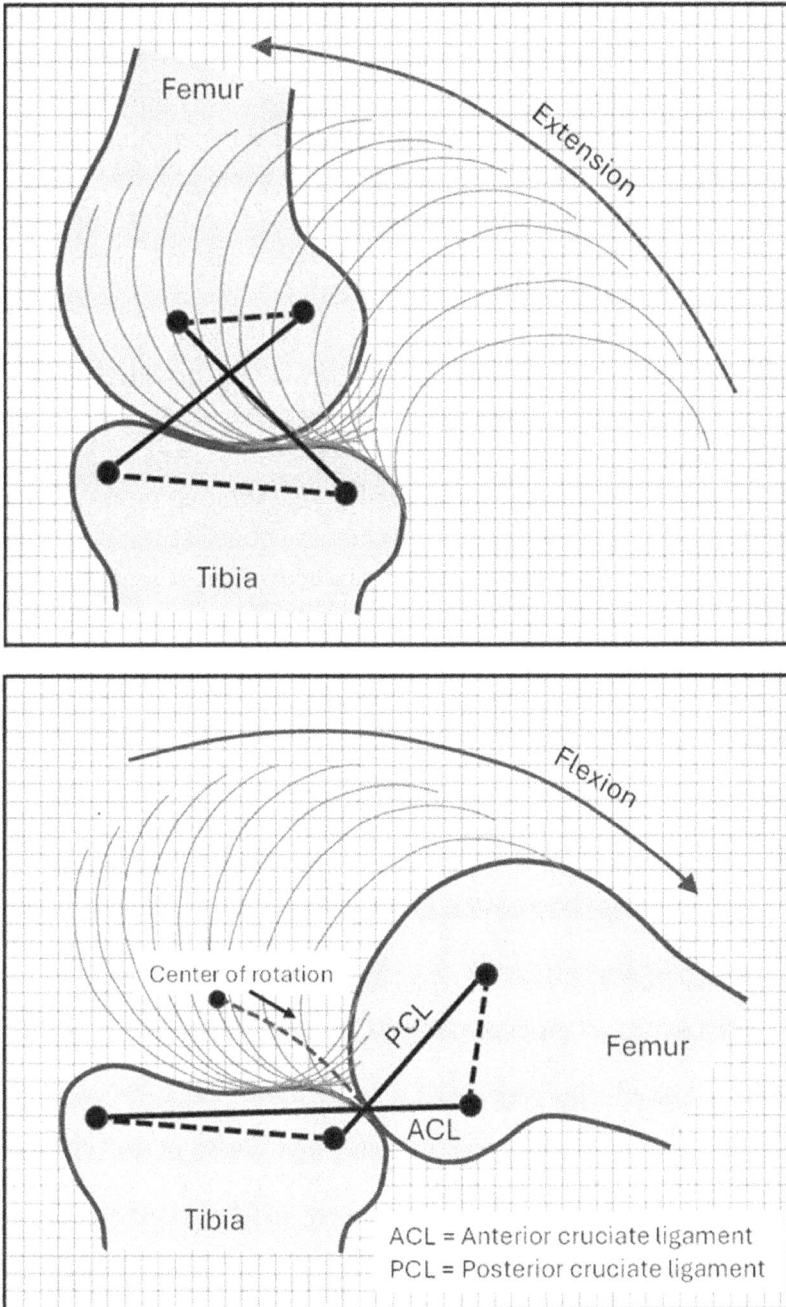

Figure 2.2. Analysis of the knee joint's linkage mechanism, carried out at Bristol University.

say "exactly," that's no exaggeration. There is little room for error in this system. If the curved profiles of the bones do not follow exact paths, the cruciate ligaments will become either too loose or too tight. My research team became acutely aware of this fine-tuning when we constructed artificial knee joints out of metal and plastic parts and discovered firsthand how difficult it was to match the linkage mechanism with the curved surfaces of the bones precisely enough for it to function properly.

Most of my researchers who worked on the knee joint were, at the end of the process, convinced the joint was intentionally designed. They could see the overwhelming evidence as they studied the details of the design.

The four-bar linkage illustrates how engineering principles are needed to understand biological systems. Exposure to these principles represents an untapped opportunity in many biology departments. I have often given crash courses in linkage mechanisms to biologists to help them understand the knee. Every anatomy and physiology department would do well to have at least one engineer on retainer to help elucidate such matters.

The Biomechanics of the Multifunctioning Meniscus

The knee has two thin, crescent-shaped pieces of fibrous cartilage known as menisci, located between the femur and tibia bones (Figure 2.1, front view) such that they surround the two condyles of the femur bone. These menisci used to be considered useless embryonic remnants but are now known to be crucial parts in the knees and are recognized as marvels of engineering.

The menisci help keep the knees aligned during flexion and extension by moving together with the condyles of the femur bone. Movement of the menisci is possible because they are cleverly anchored to the tibia by ligaments only on the outer edges of the crescent-shaped rings and not on the inner edges.

The menisci also guide the tibia during axial rotation relative to the femur bone. When the knee is in a position of flexion between around 30 and 90 degrees, the tibia is able to rotate axially relative to the femur over a range of about 60 degrees. A lesser range can be achieved outside of the 30–90 flexion range.

The menisci have several other functions besides guiding the knee in joint movement. These include load distribution, shock absorption, joint nutrition, lubrication, and aiding proprioception, the latter being the body's ability to sense its position and movements. These multiple and complex functions are possible due to the menisci's very specialized geometry and material. During activities like running, axial forces in the knee joint compress the meniscus, and its wedge shape translates the vertical compressive forces to what are known as "horizontal hoop stresses," which helps absorb the shock. The contact area between the two bones is also greatly increased, thus reducing contact stresses.[7]

The Biomechanics of Knee Locking

Standing comes so easily to healthy adults that it is taken for granted. But it is easy thanks only to some seriously clever engineering in

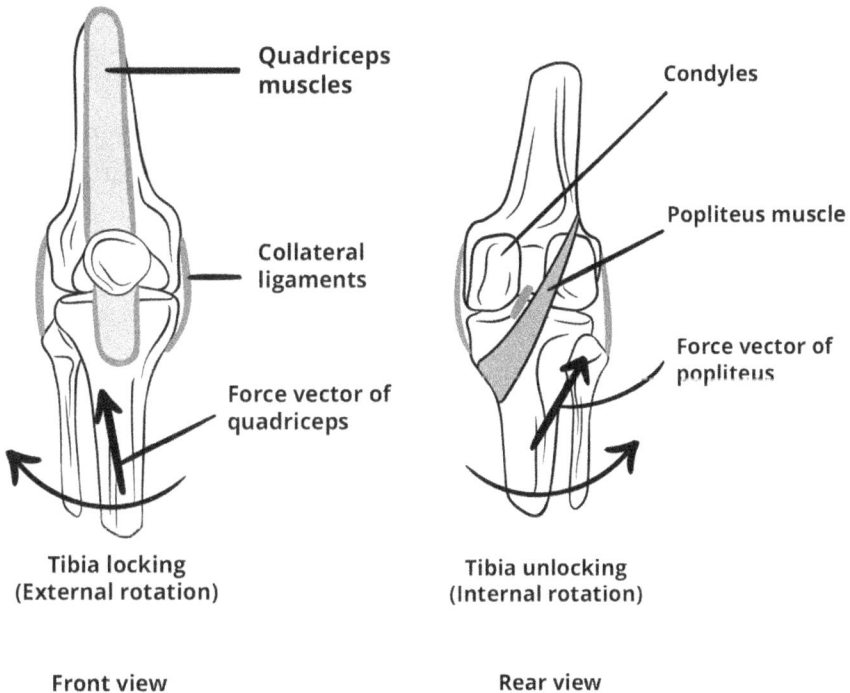

Quadriceps muscles

Condyles

Collateral ligaments

Popliteus muscle

Force vector of quadriceps

Force vector of popliteus

Tibia locking
(External rotation)

Tibia unlocking
(Internal rotation)

Front view

Rear view

Figure 2.3. Schematic of knee-locking function, front and rear views.

the knees and feet, including engineering that allows the knees to automatically lock when in the standing position, affording greater comfort and saving energy.

The knee locks in full extension by maximizing the outward rotation of the tibia. (See Figure 2.3, front view.) As the knee reaches full extension, external rotation is caused by the geometry of the condyles and by the lateral pull component of the quadriceps. In the locked position the cruciate ligaments also become tight. To unlock the joint, the tibia is made to rotate internally by flexion muscles, especially the popliteus muscle behind the knee.[8] (See Figure 2.3, rear view.) Our knees undergo this complex reconfiguration without us even being aware of it. The knee-locking function is sometimes called the screw-home mechanism.

Using the Knee Joint for Bioinspired Designs

Engineers, in awe of the design of the human knee joint, are keen to copy its mechanics as much as possible, not just for prosthetic knees but also for other engineering applications such as robotics. All of my prosthetic and knee-joint designs copy the four-bar linkage system found in the knee joint.[9]

Researchers at the Bristol Robotics Lab have produced a robotic knee that copies the cam profile of the tibia and the cruciate ligaments.[10] They have concluded that the best way to produce a compact joint with high performance, such as high mechanical advantage, is to copy the knee joint. They are far from outliers. Many researchers have concluded that copying the knee joint is the best strategy for producing better prosthetics and robots.

Another example of a bioinspired knee joint is one by researchers at Oita University in Japan. Their bioinspired design copies the condylar surfaces and the moving center of rotation using a pin and slot.[11]

While working on projects for the European Space Agency (ESA), I developed a multifunctioning lockable gearbox inspired by the knee joint's screw-home mechanism. This was used on the robotic arm of ESA's METOP satellite launched in 2019.

Figure 2.4 shows a robotic knee joint developed by my research team at Bristol University, which successfully replicates the four-bar mechanism of the human knee joint. This affords the joint a large range of motion in flexion and extension. Our artificial knee differed from the human knee in that we put the cruciate ligaments on the outside of the knee rather than at the center. The reason is that it makes assembly much more feasible. The tradeoff is that cruciate ligaments are exposed and more vulnerable, but given the constraints of current engineering technology, this was a compromise we had to make. The human knee, of course, manages to tuck the cruciate ligaments safely inside, never mind the greater demands this poses for assembly—one more instance of ultimate design in human anatomy.

I remember asking my PhD student to do a parametric study of the knee to work out the optimal design. Central to this was working out the equations that defined aspects of performance such as mechanical advantage, sliding ratio, and range of motion. I had realized that with these equations we could search the solution space for the overall optimal design. So I asked my student to work out a few thousand solutions so that we could populate some parametric graphs.

After developing an automated way of doing the calculations, we published a lengthy paper in the *ASME Journal of Mechanical Design*. Our work showed that the human knee joint is optimally designed for the multiple functions of the knee.[12] This is an example of published research that contradicts evolutionists' claims of suboptimal design in the knee.

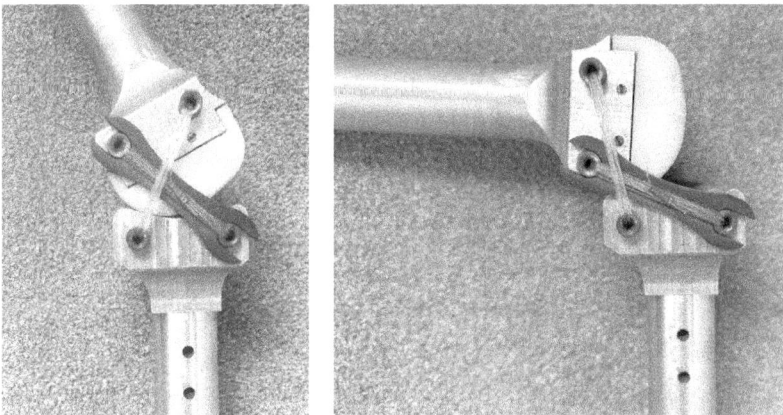

Figure 2.4. Robot knee joint developed at Bristol University.

Answering Claims of Bad Design

Despite the high praise scientists and engineers give the human knee joint, Lents, recall, insists the knee is a bad design due to the supposed incomplete adaptation from a knuckle-walking quadruped ancestor. In particular, he claims that ACL injuries are common because the ligaments are too small.[13]

Lents's claim that ACL injuries are common is misleading because ACL injuries are common only in high-impact sports like soccer and skiing. Those who take part in such sports are warned that they may sometimes experience abnormally heavy loading on the knee joint. ACL injuries are relatively rare in less dangerous sports like tennis and running.

When ACL injuries do occur in normal activities like running and jumping, the injuries are not a consequence of the original design but of other factors such as lifestyle, misuse, or disease. As with the ankle joint, Lents makes the mistake of conflating design issues and non-design issues.

Here are some general categories of things that lead to ACL injuries:

(1) Ramping up training load too quickly

If an athlete increases the training load on his knees at a recklessly fast rate, he shouldn't be surprised if he suffers an ACL injury. This is a case of misuse. It is well known that training loads should be increased gradually to allow muscles and ligaments to adapt in size.[14] If loads on the knee are gradually increased in training, then the ACL gradually adapts, growing in size and strength. Human engineers, it should be noted, are nowhere close to replicating this design feature with artificial knees.

(2) Poor childhood lifestyle

Studies have shown that when children are not active, then joints such as knee joints are more liable to experience problems later in life.[15] Childhood exercises such as running and climbing strengthen joints by developing the muscles and ligaments. Without adequate exercise in childhood, these parts are not fully developed.

(3) Being overweight

Being overweight is one of the most common reasons for knee problems in modern society.[16] Overweight persons must restrict their sporting activities to avoid overloading their knees. In the US, more than two-thirds of adults are overweight and one in three are obese. This is a problem for the knees, because while the ligaments can adapt in size for increased sporting loads, the ligaments cannot adapt so well for increased weight. If a person is ten pounds overweight, this may not sound like much, but it results in an extra sixty pounds of force on each knee joint during running, an enormous strain. This is why obese persons are twenty times more likely to require a knee replacement than individuals who are not overweight.[17] In the US, the percentage of adults who are obese has more than tripled over the last sixty years, so it is no surprise that ACL and other knee injuries have become more common.

(4) Disease

Diseases like arthritis also compromise knee performance. Some of these diseases are aging-related, but there is evidence that an unhealthy Western diet high in sugar and other processed carbs exacerbates diseases like arthritis. Those with such diseases have to reduce the usage and loading on their knees to avoid injuries.

If you have some combination of the above risk factors—i.e., overweight, a sedentary childhood, unhealthy diet, participation in dangerous sports—then your chances of developing knee problems increase dramatically.

The fact that ACL injuries were rare in past generations tells us that a major factor behind such injuries is the modern lifestyle. Misusing the knee is analogous to misusing a human-engineered product. If you were to carry lots of heavy weights in the trunk of a well-engineered sports car all the time, your suspension would wear out far faster than if you did not carry weight the car wasn't designed for. In this case, the problem is not the car design but the car's misuse. In the same way, most knee problems today are due to misuse and abuse of the human body.

A Rolls-Royce of Joints

The knee joint is an ingenious design with its rolling action, four-bar linkage mechanism, and precise geometry. There are many scientific publications detailing the fine-tuning of design in the knee joint. Lents missed all this, likely due to his prior commitment to the evolutionary paradigm and his consequent failure to follow the scientific evidence.

His lack of an engineering background didn't help either. The knee's linkage mechanism requires advanced engineering knowledge to fully understand. Its four-bar linkage again illustrates how engineering knowledge is important for a proper understanding of, and appreciation for, many biological systems. When engineers study the human knee joint in detail, they readily see that they have before them a case of ultimate engineering. Yes, knees wear out and are not invulnerable to misuse. But when a healthy knee is looked after, it is amazing how well it performs.

3. THE WRIST JOINT

Two Views of the Wrist Joint

Researchers have given high praise for the design of the human wrist. One medical website states that the wrist joint has "a remarkable range of movement."[1] In a scientific paper for a leading robotics conference, Neil M. Bajaj and his colleagues at Yale University reviewed sixty-two of the world's best prosthetic wrists. Their conclusion: "The most apparent observation is the scarcity of 3 Degree of Freedom compact powered wrist devices." And "scarcity" is putting it mildly. The authors concluded that no prosthetic wrist could match the compactness and functionality of the human wrist. In comparison to it, prosthetic wrists were bulky with poor functionality.[2]

Additionally, in 2024 I published a state-of-the-art review paper on the multifunctionality of human and animal joints.[3] The paper shows that multifunctionality is a key reason for the extreme levels of agility found in animals and recommends that robot designers emulate this feature to produce better robots. One of the paper's main case studies of brilliant multifunctionality in nature is that of the human wrist joint. The paper was well received, gaining acceptance for publication in the well-ranked journal *Biomimetics* and chosen as the cover story and lead paper out of seventy-two for that issue, further evidence that such findings are being embraced by the mainstream of experts in the field.

Despite the fact that the wrist joint is superior to any prosthetic wrist that engineers have produced, some critics insist that it is a terrible design. "The wrist bones and ankle bones," writes Nathan Lents, "are the most obnoxious examples of bones for which we have no use."[4]

Here again he has been drawn into serious scientific error by a paradigm that anticipates poor design, with the evolutionary process assumed to have clumsily evolved a knuckle-walking quadruped into a bipedal human. I fully agree with Lents that evolutionary theory predicts the wrist should be a poor design. But all scientific research points in the opposite direction, to the wrist being a brilliant design and far beyond anything human engineers have produced.

The Incredible Multifunctioning Wrist

We so easily take the wrist for granted, but it is one of the key reasons for the extraordinary dexterity of the hand. We need the wrist joint for activities such as getting dressed, washing up, playing most ball sports, and playing musical instruments. It also made our forebears uniquely skilled at farming, building, and hunting.

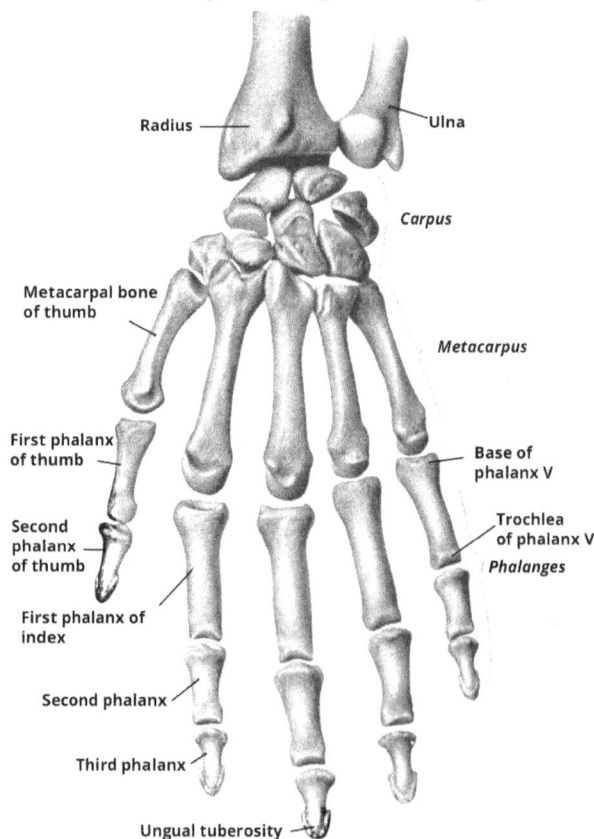

Figure 3.1a. Anatomy of the wrist bones (carpus) and finger bones.

To perform the complex movements in these activities, the wrist needs to be able to rotate in three perpendicular directions. (1) The wrist can rotate axially for actions like turning a door handle or turning a screwdriver. Pronation of the forearm rotates the palm of the hand down and supination rotates the palm up. (2) Forward bending (moving the palm of the hand toward the forearm) is called flexion; backward bending (moving the back of the hand toward the forearm) is called extension. (3) The wrist also can flex to either side—abduction (away from the midline) to the thumb side and adduction (toward the midline) to the pinky (little finger) side—actions used in waving and washing. Some actions, like throwing a curveball in baseball, involve a combination of the three joint rotations.

The wrist is what engineers call a universal joint because it can move in different perpendicular directions. Engineers use universal

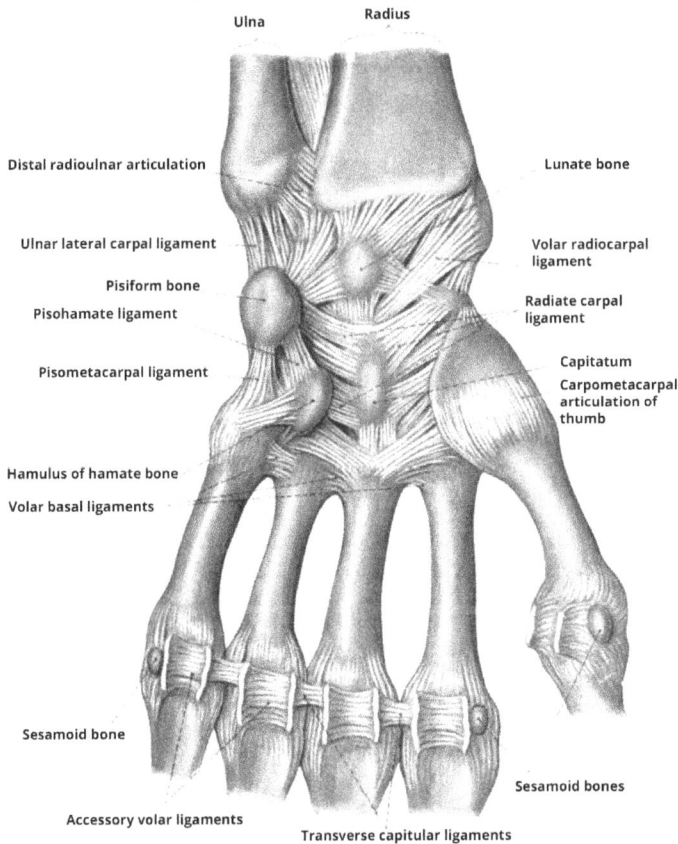

Ulna
Radius

Distal radioulnar articulation

Lunate bone

Ulnar lateral carpal ligament

Volar radiocarpal ligament

Pisiform bone
Pisohamate ligament

Radiate carpal ligament

Pisometacarpal ligament

Capitatum
Carpometacarpal articulation of thumb

Hamulus of hamate bone
Volar basal ligaments

Sesamoid bone

Sesamoid bones

Accessory volar ligaments
Transverse capitular ligaments

Figure 3.1b. The main ligaments of the wrist.

joints for advanced applications such as drive systems and instruments like wrenches (called *spanners* where I'm from) that have to work in awkward spaces. They understand that universal joints are complex mechanisms that require precision engineering.

The wrist, besides being able to move in three different ways, must also be strong enough to withstand the large loads that sometimes come through the hand. For example, when doing a push-up, much of the body weight passes through the wrists.

Figure 3.1 gives a glimpse of the wrist's complexity, showing the bones and ligaments. (See also Figure 4.1.) There are eight bones in the wrist, each with precise interfaces to the surrounding bones. Notice how the eight bones are arranged approximately in two rows of four bones each. This arrangement is the key to multifunctioning in the joint, as we shall see in the following sections. There are four main ligaments but more than thirty ligaments in total holding the bones together. Every ligament has a precise function. And there are six main muscles that move the wrist. These are located in the forearm.

The wrist also integrates many other critical structures. There are around twenty tendons passing through the wrist, and these must be given freedom to move lengthwise, without friction, when the fingers are being moved. There are three main nerves that pass through the wrist joint: the ulnar, median, and radial nerves. The wrist also must accommodate the radial and ulnar arteries, the main blood supply to the hand. Additional blood vessels traverse the back of the wrist, supplying blood to the back side of the hand. There are also bursae for lubricating the tendon movements.

Despite these many parts, the wrist is remarkably compact. It is so slender that it is a good place to feel the pulse in the radial artery. At the same time, the wrist is remarkably robust. If we use the wrist with care, it is capable of working for an eighty-year lifespan, performing millions of movements. The volume of the eight wrist bones is small (for an adult typically about that of a golf ball), and yet the amount of functionality packed into the wrist is astounding.

This is why engineers are in awe of the wrist joint.

Biomechanics of Wrist Rotation

Figure 3.2 illustrates how the wrist rotates axially in pronation-supination. The wrist forms a joint with the ulna and radius such that the hand can be rotated as shown. The forearm actually forms a type of four-bar linkage system, this one known as a crossover four-bar mechanism.

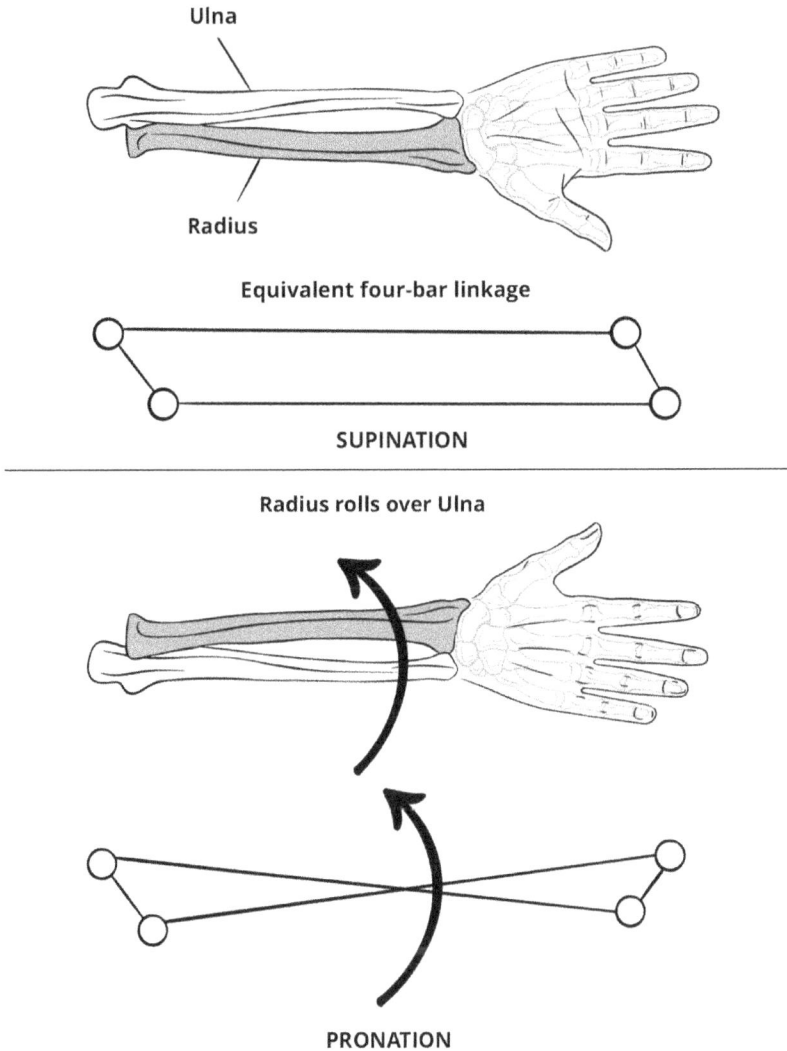

Figure 3.2. Biomechanics of wrist rotation. The long links are formed by the ulna and radius bones. The short ones are virtual links produced by the way the bones are joined by ligaments. The linkage mechanism enables the ulna and radius to rotate around each other in a precise way.

One challenge with this linkage system is that it can limit the elbow joint's range of movement. However, the ulna bone has an ingenious feature, a special notch (called the trochlear notch) that lends the elbow joint virtually full range of motion for any angle of pronation in the wrist.

All this illustrates, again, why engineering principles are essential for understanding and appreciating important aspects of biology. They are needed but often sorely absent. This masterful four-bar linkage system produces a stable joint with smooth motion, and yet the vast majority of biology books that describe the joint make no mention of the four-bar linkage.

Notice also that, as with all four-bar mechanisms, the forearm is irreducibly complex: All four bars are needed for the system to work. When you have only two bars, all you have is a simple hinge. When you have three bars, the result is a rigid structure that can no longer move. To form a four-bar mechanism requires advanced planning of the four linkages and their joints.

The Biomechanics of Adduction-Abduction

Figure 3.3 shows the layout of the eight bones during adduction of the right hand (moving sideways toward the pinky side of the hand). During adduction and abduction, the main sliding movement occurs at the radiocarpal joint, between the wrist (carpus) and radius bone. However, a smaller but important amount of sliding also occurs at the midcarpal joint, between the two rows of four bones. This extra play at the midcarpal joint gives the wrist a large range of motion.

The presence of two sliding joints (radiocarpal and midcarpal) raises the potential problem of having two different centers of rotation. The only way to avoid this is for both joints to produce the identical center of rotation. However, to achieve this the radii of the two joints must be fine-tuned to exactly the right ratio. The common center of rotation is shown in the diagram, with a circular symbol.

To have the same center of rotation, the radius of the midcarpal joint (MCJ) must have a radius smaller than that of the radiocarpal joint (RCJ) and by exactly the right amount. And this is precisely what the wrist has, as shown in Figure 3.3.[5]

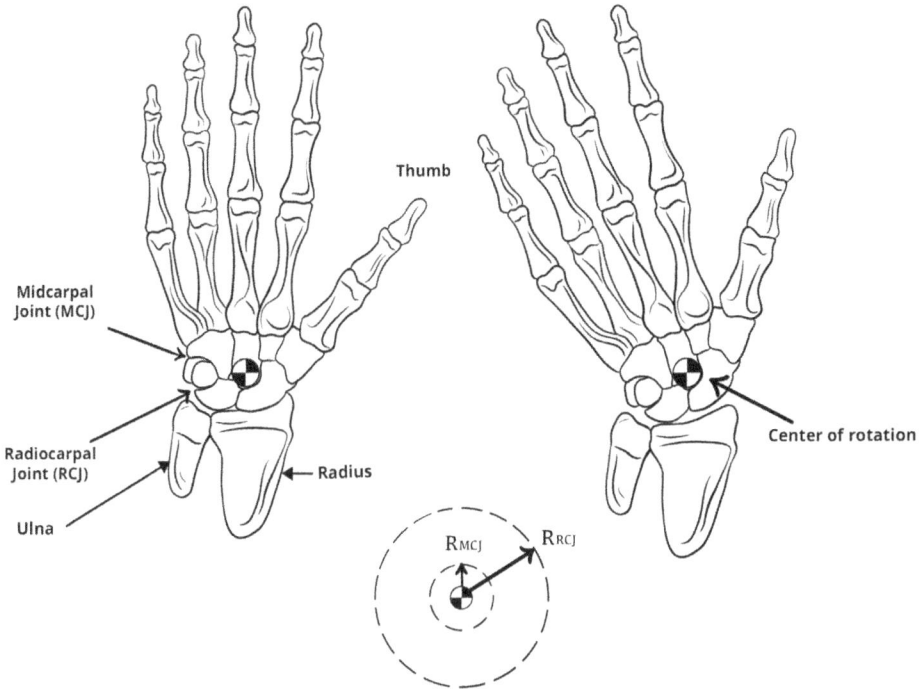

Figure 3.3. Adduction of right hand (palmar/palm view), plus a diagram showing the common center of rotation. The radiocarpal joint extends through the width of the wrist, between the ends of the radius and ulna bones and the first row of four wrist bones. The midcarpal joint also extends through the width of the wrist, running between the two rows of four wrist bones.

The Biomechanics of Flexion-Extension

Figure 3.4 shows the biomechanics of flexion and extension in the wrist. A notable design feature is how the bones are curved at the midcarpal joint to allow flexion-extension to occur. This means the bones are curved in two directions at the midcarpal joint—one curve for abduction, one for flexion. That involves a very complex curved geometry in three dimensions. Engineers refer to this design as an ellipsoidal joint.

As with abduction-adduction, during flexion-extension sliding movements occur at both the radiocarpal and midcarpal joints. As before, the two sliding joints (radiocarpal and midcarpal) create the potential for two centers of rotation and chaotic motion unless the joints are fine-tuned to have a common center of rotation. This common center of rotation is shown in Figure 3.4, with a circular symbol. To have the same center of rotation, the midcarpal joint must have a radius smaller than the radiocarpal joint and by exactly the right amount. And once again, this is precisely what the wrist has.[6] What engineers find especially remarkable is that fine-tuning happens in two different directions.

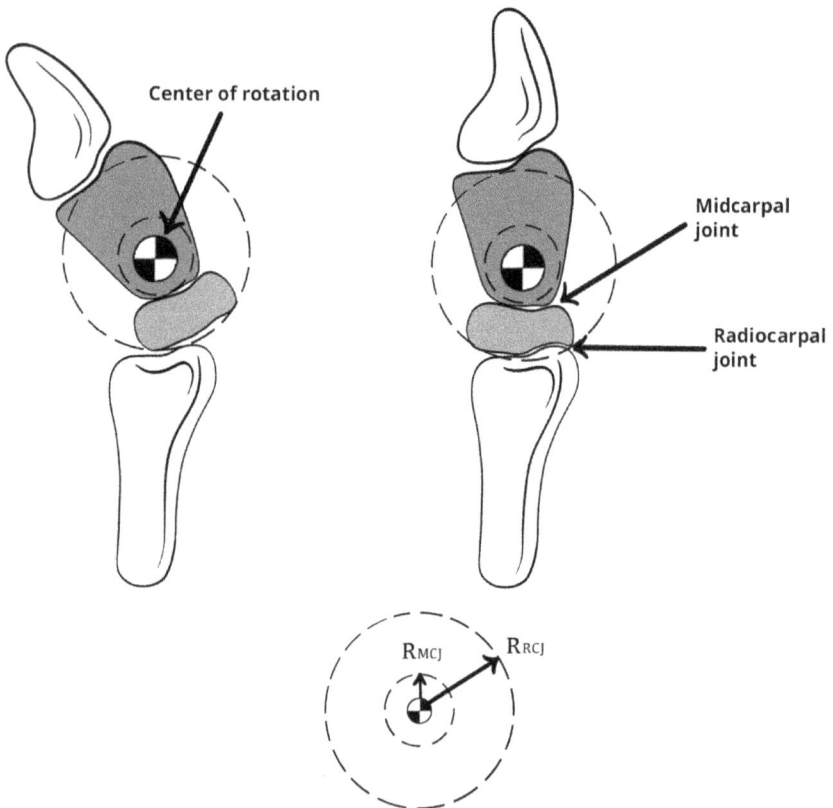

Figure 3.4. Wrist flexion-extension (side view), plus a diagram showing the common center of rotation.

Carpal Tunnel Function

The carpal tunnel is formed from an arch made from six of the eight wrist bones. The six bones include all four from the top (distal) row and two bones from the bottom (proximal) row, shaded in Figure 3.5. There is also a (transverse) carpal ligament on the palm side of the hand that forms a complete tunnel, as seen in Figure 4.1. This arch has the important job of providing a protective tunnel for delicate tendons, nerves, and blood vessels.

Incredibly, even though the eight wrist bones are fine-tuned to produce two joints, the same bones are perfect for forming a carpal arch. In fact, the wrist actually forms other arches (together with the fingers) that give the palm of the hand a naturally concave shape, ideal for grasping round objects. This means the wrist forms multiple arches in different directions. The astonishing degree of precision engineering involved in achieving this multifunctionality is nothing short of a design masterclass.

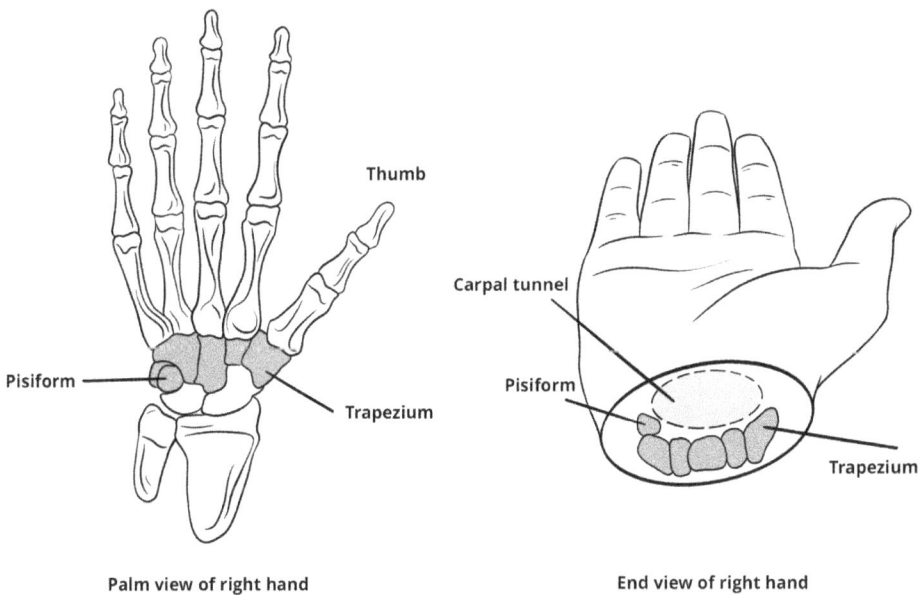

Thumb

Carpal tunnel

Pisiform

Trapezium

Pisiform

Trapezium

Palm view of right hand

End view of right hand

Figure 3.5. Carpal tunnel function.

The Load Transfer Function

In addition to serving all the functions described above, the wrist bones also must be able to transfer high loads from the fingers to the arm. (See Figure 3.6.) When doing a handstand or pushing a heavy cart, considerable weight goes through the wrist joint, so the joint must be strong. Transferring loads is especially challenging because there are five fingers but only two forearm bones. Structural engineers know it is notoriously difficult to transfer loads between two sides that have different numbers of elements. But thanks to a masterful design, the carpal bones of the wrist readily handle the task.

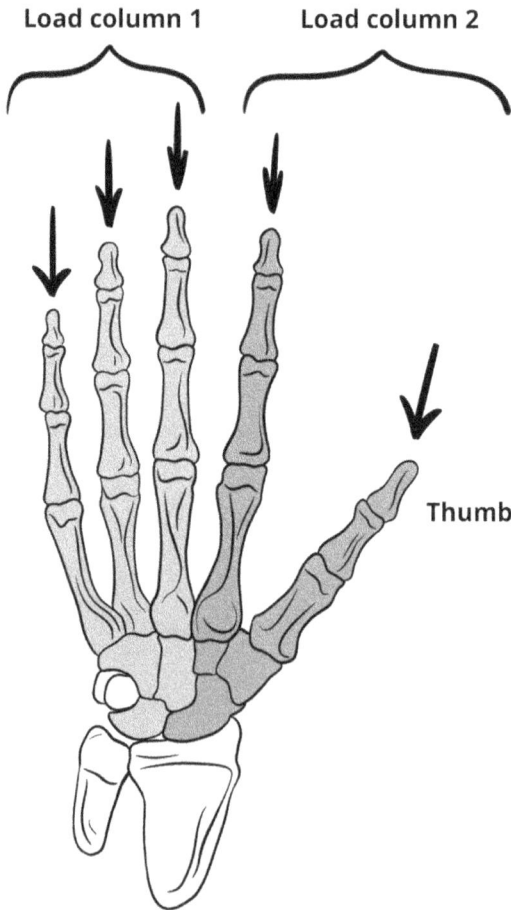

Figure 3.6. Load transfer function.

Notice in Figure 3.6 how the carpal bones produce two load paths that bring the loads from the fingers into the center of the radiocarpal joint. Also notice how the carpal bones line up precisely with the thumb, index finger, and middle finger. This produces an efficient flow of forces through the hand. Notice, too, how one carpal bone lines up with both the fourth and fifth finger. This makes sense because the loads through those fingers are relatively low compared to the first three fingers.

In my 2024 journal paper on multifunctioning, I presented a table to summarize how the same set of eight bones produces four major functions in the wrist, with each function highly optimized. It bears repeating: This degree of multifunctioning is astonishing. It's part of why many engineers call the wrist one of the pinnacles of mechanical design.

Answering Claims of Bad Design

The human wrist, says Lents, "is way more complicated than it needs to be." Its bones, he says, are "like a pile of rocks—which is about how useful they are to anyone. Collectively, these wrist bones are helpful, but they don't really do anything individually. They sort of sit there when you move your hand…. No sane engineer would design a joint with so many individual moving parts. It clutters up the space and restricts the range of motion."[7]

It is hard to convey the magnitude of Lents's error here. In reality, the eight wrist bones are a particularly brilliant example of how one set of bones can perform multiple complex functions. To say that having this many wrist bones is useless is akin to someone unaware of smartphone versatility looking inside one and, baffled by its complexity, declaring that the phone is stuffed with a useless plentitude of parts. Insisting that "no sane engineer would design a joint with so many individual moving parts" demonstrates a clear lack of engineering knowledge. Engineers are in awe of the wrist joint and would love to copy it if they had the materials and technology to do so. Lents also has chosen the wrong design to make an accusation of wasting space. The wrist joint is recognized as one of the most compact joints in the world.

Of course, there can be medical problems with the wrist joint, but these are due not to any supposed flaws in its original design but instead to misuse, accidents, and genetic disorders. Take carpal tunnel syndrome, for example, one of the most common wrist disorders in modern society. Carpal tunnel syndrome involves swelling or damage in the wrist, which puts pressure on the median nerve. Symptoms of carpal tunnel syndrome can include wrist pain, along with tingling, numbness, and weakness of the fingers and hand. The great majority of people suffering from it are involved in repetitive activities such as typing or operating vibrating tools. Such activities cause tendons to be heavily loaded in the same directions for long hours, day after day. Eventually there is an inevitable fatiguing effect and a degradation of the wrist. Before the industrial age, such disorders were much less common, showing that modern technology has been a key factor in repetitive strain injuries.

We might be able to imagine a redesigned wrist that is better for repetitive activities, but in doing so we would be trading away other functions of the wrist that it was designed for. No engineered system can handle every possible use for endless periods of time. The same principle applies to human joints. The question is, How well does the human-wrist design allow for a wide range of functions, and how well does it manage inevitable trade-offs? The answer: vastly better than the best human engineers have thus far managed.

The Rolex of Joints

The wrist joint is recognized by the biomechanics community as ultimate engineering. The bones in the wrist are like the gears in a Rolex watch in the way they mesh beautifully. In fact, the precision engineering of the wrist joint surpasses that of a Rolex watch. Lents's manifestly false claims of bad wrist design show the damaging effect of imposing an evolutionary paradigm on the human body. The wrist joint shouts not just intentional design but the kind of masterful design one could only expect from an engineering genius.

4. FINGERS

Two Views of Human Fingers

There are nine muscles that control the tendons of the thumb. For such a small structure to have so many actuators is astounding. I have carried out research on human fingers and thumbs, and from this experience I can testify that human fingers exemplify a design virtuosity that human engineers are nowhere close to matching, much less exceeding.

I was part of a team that developed an exoskeleton hand used by stroke patients. We had to design a robotic device that could fit over the hand of a stroke patient so as to strengthen the hand enough to allow the person to carry out tasks like gripping a jar.[1] During our project we discovered that the human hand was extraordinarily dexterous. We also learned that robots can only simulate a limited number of human hand movements. Despite the inferiority of human technology, it is still very fulfilling to create medical aids that can transform the life of people with disabilities. There is also a poetic satisfaction in reflecting that the skill of our hands was used to create the exoskeleton.

The brilliant design of fingers makes good sense if we understand them to have been intelligently designed. But that same brilliant design poses a major challenge to Darwinism. According to evolutionary theory fingers should not be masterfully designed, this due to their having evolved from knuckle-walking apes and because Darwinian gradualism should yield a finger-hand complex that is only good enough for survival. Yet the capabilities of the human hand far exceed what are needed for primitive survival.[2] According to evolutionary theory, the hand has evolved for primitive survival tasks like fist fighting,[3] clubbing,[4] basic stone tool making, eating, and relatively simple food preparation. With such a relatively modest set of requisite abilities, our fingers should not be capable of extremes of fine skill, but they are.

The Extraordinary Dexterity of Human Fingers

Human fingers are extremely dexterous. When learning high-dexterity skills in childhood, our brains memorize the groups of muscle movements involved till the skills are second nature. Here are a few things human hands can manage, some of them within the reach of ordinary people, some of them outlier examples:

- Make multiple grips when tying shoelaces.
- Manipulate delicate objects with precise forces, allowing us to do things like break an egg and dump its contents into a frying pan while maintaining hold of the shell. (Most people can be taught to do this one-handed.)
- Make two stitches per second involving several complex movements.
- Play an 88-note chromatic scale with one hand in six seconds.
- Thumb-type at about ten characters per second on a smartphone.
- Rotate the various parts of a Rubik's cube at ten rotations per second.

The Rubik's cube involves making turns in three dimensions in order to line up colors. The player must work out the correct sequence of turns as well as make the turns. In 2023 a contestant, Max Park, broke the world record, solving the cube in 3.13 seconds using thirty-three rotations. That works out to a furious 10.5 turns per second.

Many advanced typists regularly type ten characters per second, with the very fastest typists able to hit speeds of fifteen or more characters per second.

Human fingertips also have a sense of touch that has astonished scientists. Research has shown that our fingertips can feel a ridge of just thirteen nanometers, which is thirteen millionths of one millimeter.[5] The distance is so small that humans can feel bumps at the scale of individual molecules of matter. This sense of touch is possible due to thousands of precision pressure sensors in each fingertip.

All this is why engineers are in awe of the design of human fingers.

Even the best robotic hands struggle to perform many daily hand-related tasks that we take for granted. Yet even children can do these complex movements from an early age. Such capabilities are due to some mind-blowing design features, as we will see below.

Finger Tendons: The Ultimate Pulley System

The great dexterity of the fingers is due in part to the presence of around twenty-seven joints and thirty-four muscle-tendon units in each hand. Figure 4.1 gives a glimpse of the amazing network of tendons on the palm of the hand. The diagram shows each finger at different levels of dissection to reveal different features. There are two tendons on the palm side of each finger—the flexor profundus digitorum and the flexor superficialis digitorum (formerly known as the flexor digitorum sublimis). Notice how the flexor superficialis tendon on the ring finger splits into two so that the profundus flexor tendon can pass through it. And notice on the index and middle fingers how sheaths protect the tendon.

Another important feature shown in Figure 4.1 is how several of the tendons pass through the wrist into the lower arm. In engineering terms, the fingers are precision cable-pulley systems where the tendon is the cable and the joints and guides in the fingers are the pulleys.

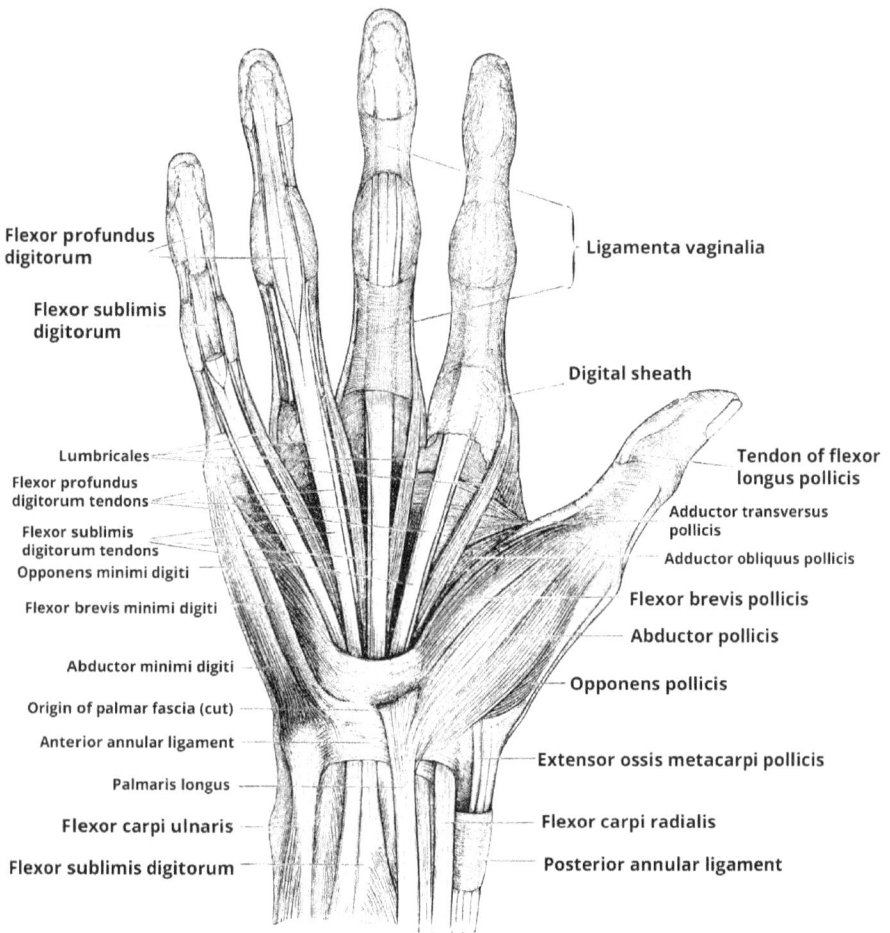

Figure 4.1. Tendons and muscles in the palm of the right hand.

A key design feature of our fingers is that around half their muscles are in the forearm (extrinsic muscles) and the other half are in the palm (intrinsic muscles). This clever design feature results in great compactness, allowing for greater dexterity. However, to achieve such a design requires an incredible feat of engineering. The way that tendons thread their way from the forearm through the wrist and into the fingers is a marvel of mechanical design.

A common example of a human-designed cable-drive system is found on bicycle brakes. When the rider pulls the lever, a force is transferred via a long flexible cable to the brake on the wheel, a cable that runs inside a protective sheath. This design strategy is also found in the hand, although in a much more sophisticated form. The hand has various sheaths to guide, lubricate, and protect the tendons.

Cable-driven systems are also used on spacecraft to deploy solar panels. These are far more precise than those used on bicycles. I have helped design cable-driven systems for NASA and ESA and am familiar with the complexities of the design. Nevertheless, their cable systems are crude and simple compared to the tendons of the human hand.

Designed for Dexterity

The anatomy of the tendons in the index finger is shown in Figure 4.2. The index finger is the most agile of the four fingers and has seven muscles controlling its movement. Figure 4.2 shows how each individual finger bone has tendons attached to both the top side and bottom side so that every joint can be moved in extension and flexion. This maximizes the number of possible movements in the finger and hence improves dexterity.

Achieving all those connections requires intricate, precision engineering. The long extensor tendon is particularly complex in the way it branches into three separate elements, including the lateral bands. This enables the tendon to connect to both the second and third finger bones.

Figure 4.2 (side view) shows how fingers have around nine guide tunnels. Figure 4.2 also includes a cross section of the finger, showing three tendons passing through a synovial sheath that lubricates the tendons to produce smooth motion. The reason for so many guides on the palm side of the finger is that whereas extension pulls the tendon onto the finger, flexion tends to pull the tendon away from the joints.

The tendons are tiny—just a few millimeters wide and 1 mm deep—and yet they can apply large forces to the finger. As well as aiding flexion and extension, the interossei muscle-tendon units can also move the fingers sideways (called abduction and adduction).

Additional parts not shown in the diagram include skin, blood vessels, nerves, three fat pads on the palmar side of each bone, and three sets of ligaments that hold the three joints together. To package so much onto a single finger is a breathtaking feat of engineering.

The impressive dexterity of the fingers will be illustrated in the following sections.

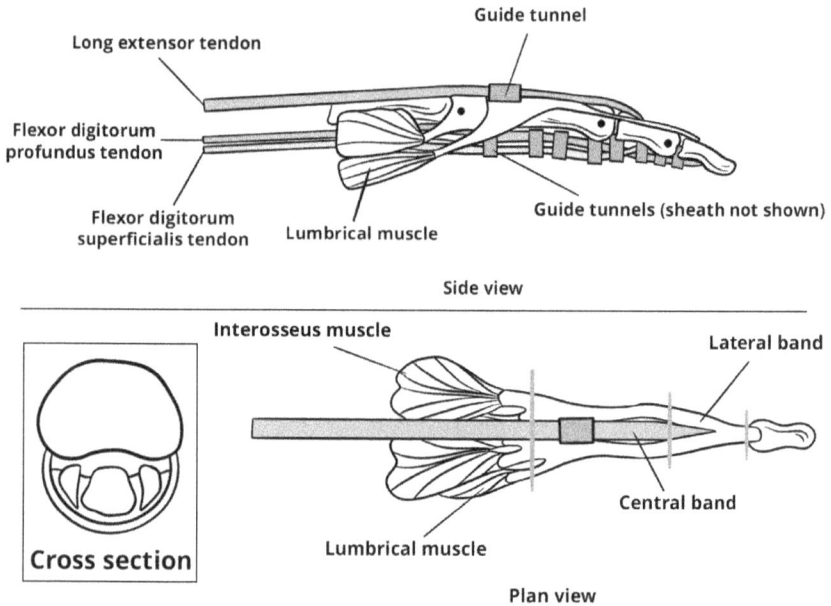

Figure 4.2. Anatomy of the index finger.

Biomechanics of Complex Finger Movements

Most finger movements require the pull of multiple tendons and involve antagonistic muscle forces (simultaneous flexion and extension forces). The combination is crucial for fine control of joint movement. The muscles located in the hand (intrinsic muscles) are mainly responsible for fine control of the fingers.

Full flexion is handled by the flexion tendons. Full extension is handled by the extensor tendons. Half flexion-extension is handled mainly by the superficialis flexor and the extensor tendons. When you flex and extend your fingers, you may be able to see muscles move in your lower arm. Full flexion is used for power grips such as holding an ax or tennis racket. During full flexion, the moment arm of the flexor tendon at the knuckle increases as the flexion angle increases, giving

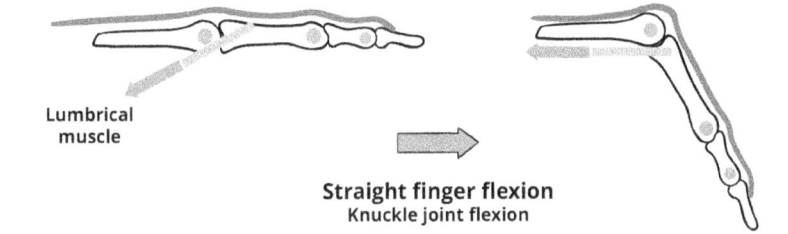

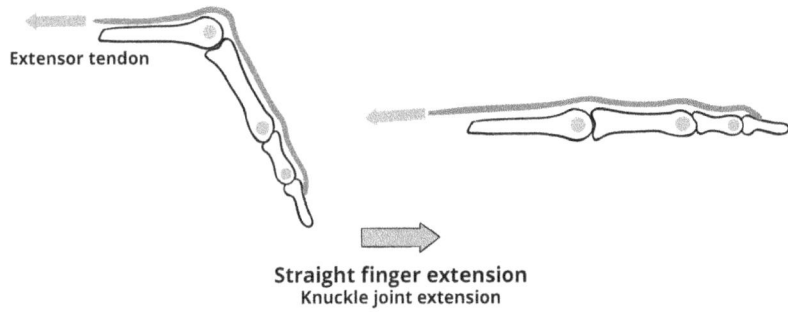

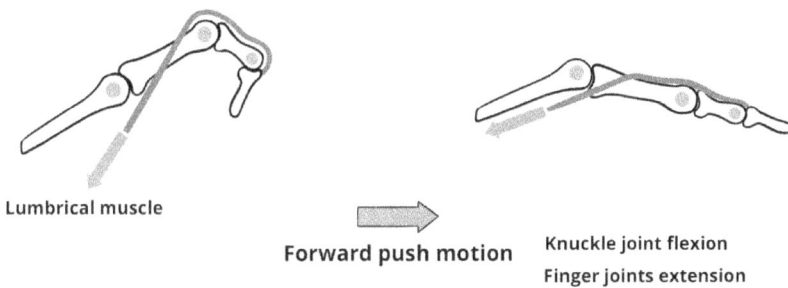

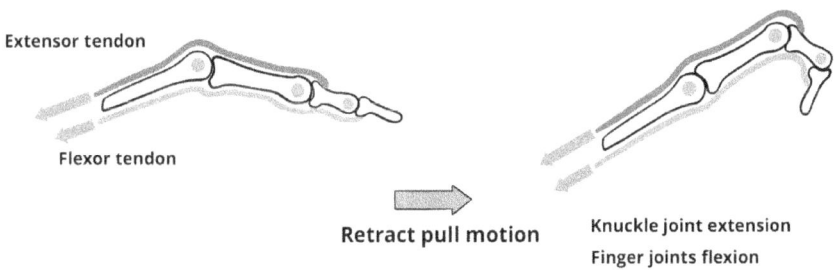

Figure 4.3. Biomechanics of complex finger movements.

the finger a strong flexion force and hence a stronger grip. This is just one of many ways the design of the finger is fine-tuned.

Babies have a palmar grasp reflex so that by around three months they can fully flex their fingers and grasp objects. When holding babies, I love to tickle their palms to test the palmar reflex. I also find it fascinating to observe how a baby gradually learns hand-eye coordination during the first few years of life.

Figure 4.3 illustrates some of the main tendon movements for certain complex finger movements. These can take children a long time to learn. Knuckle flexion-extensions involve the finger being kept straight while the finger is rotated around the knuckle joint. The lumbrical muscle is cleverly positioned at an angle so that it flexes at the knuckle while remaining extended at the two finger joints, allowing the lumbricals to flex a straight finger. To extend a straight finger, the extensor tendons pull the finger back.

The forwards-backwards typing motion is a particularly complex finger movement. The lumbrical tendon is key to the forward movement because it flexes the knuckle joint while extending the second and third finger joints. Pulling the finger back requires different tendons to be pulled simultaneously. Despite the extreme complexity of the typing action, once a person has mastered typing, the brain can instruct the precision movements of a group of muscle-tendon units without effort.

Biomechanics of Abduction and Circumduction

There are two sideways finger movements mainly performed by the two interossei muscles on each finger. The dorsal interossei performs abduction, spreading the fingers apart. The palmar interossei performs adduction, bringing the fingers back together.

Circumduction involves the end of the finger moving through a circular movement. Most people find this very simple to do, but it is really a very complex dynamic movement, so much so that one could carry out a master's degree in biomechanical engineering just studying this one movement. It requires a precisely timed sequence of different muscle movements. The circle's first quadrant involves abduction plus extension; the second quadrant, adduction plus extension; the third, adduction plus flexion; and the fourth, abduction plus flexion. Then the cycle repeats.

Circumduction involves precise changes in all the tendon speeds over time, with different tendons moving in different directions. In order to make a constant and smooth circular motion at constant speed, the tendons must gradually accelerate and decelerate in motion.

That our brains and fingers can orchestrate all this, all with very little conscious effort, is astounding.

The Biomechanics of Fine Motor Control

Fingers also have muscles fine-tuned for skillful moving. Each of the thirty-four individual muscles for the fingers of a hand contains individually controlled muscle sections called motor units. For example, the first dorsal interosseus muscle has around 140 motor units.[6] That means each hand has thousands of motor units. Each motor unit has its own nerve pathway, which means the brain can guide the fingers to create very finely tuned forces by recruiting just one (or a few) muscle units.

The fine control of finger muscles that this affords helps explain why humans have such a delicate touch in areas like music, art, cooking, and surgery.

The design of the human brain also contributes to the extraordinary dexterity of our hands. The motor cortex is the part of the brain that controls the more than six hundred muscles around the body. Even though the hand muscles represent about 10 percent of the body's muscles by number (and much less by mass), fully 25 percent of the motor cortex is dedicated to controlling the hand muscles. The extra brain space for the hands is required to store the complex patterns of muscle movement like the push-pull motion used in typing.

Superior to Robot Fingers

Human fingers are not just supremely dexterous; they are also remarkably robust. They can perform millions of movements over a lifetime of eighty years, heal from occasional injuries, and remain highly capable.

Several top musicians have continued playing music at a professional level into their older age. At age ninety, German pianist Menahem Pressler played to a high standard with the Berlin Philharmonic and recorded albums of Mozart and Debussy.

Figure 4.4 shows an exoskeleton hand that my research group developed at Bristol University. It copies the cable-pulley system of the tendons in the human hand to flex and extend the fingers. We designed it to fit a particular patient who had recently suffered a stroke that severely weakened her right hand. We got to see the patient try it out, and it was very moving to see her regain the ability to hold objects by using our exoskeleton. The prosthetic device was good enough to dramatically improve her quality of life. At the same time, none of us involved in the project deluded ourselves: Our exoskeleton could only make simple hand grips, far short of what a healthy human hand could manage.

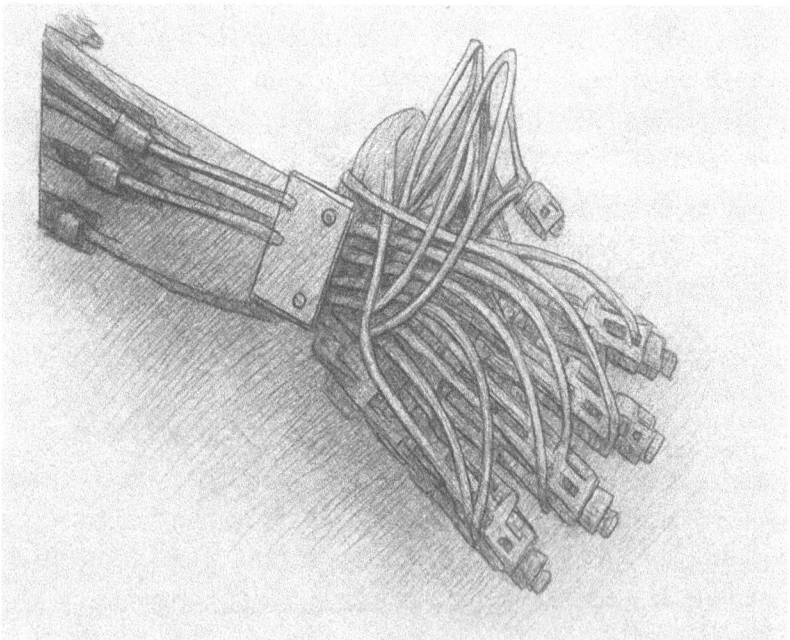

Figure 4.4. Exoskeleton hand developed at Bristol University.

This wasn't just a case of our prosthetic being behind the technology curve. In contrast to human fingers, robot fingers are, across the board, far less dexterous, far less touch sensitive, far bulkier and less robust. Comparing robot fingers to human fingers is like comparing a go-kart to a Formula 1 race car.

Evidence Points to Intelligent Design

So where does this leave us? We know robot fingers are intelligently designed. And yet we are told not to even consider the possibility that human fingers were intelligently designed and instead to trust evolutionary theory. But as we saw, there is a problem with that idea. Human hands possess a level of dexterity far beyond what is needed for survival tasks like tool-making or wielding a club, and therefore far beyond what we could expect a gradual evolutionary process guided by immediate fitness requirements to ever generate.

The level of skill that human hands demonstrate in areas such as art, music, and surgery is stunning. To be able to play fifteen notes per second with one hand, and with precision, is not needed for survival. To be able to feel a ridge of thirteen nanometers is also not needed for survival. The great skill of human hands is a case of what we will further explore later in the book: purposeful overdesign.

5. The Spine

Two Views of the Human Spine

I have written several journal papers on the optimal design of structures in engineering and nature.[1] When I study the human spinal column, I find a system optimized to provide a balance of stiffness, strength, and flexibility. In this I am merely recognizing what is now conventional wisdom among engineers familiar with the human spine, so much so that some now look to it for inspiration for future engineering structures.

The journal *New Civil Engineer*, for example, describes why engineers are using the human spine as inspiration for designing the advanced bridges of tomorrow:

> The human spine has provided the inspiration for the design of a new low-cost bridge that could improve transport infrastructure resilience.
>
> Researchers at Southampton University have developed the concept, using the human skeleton as the basis for their proposal for a durable, low maintenance, low cost bridge.
>
> Intervertebral disks in the spine provide flexibility, dissipate energy from the movements of the body and absorb and transmit forces without damaging the bones, and academics will base their new design around these concepts.[2]

These UK-based researchers, at Southampton University and Bristol University, have received more than a million dollars of funding for the effort. They hope to create flexible bridge joints inspired by those in the spinal column, with the aim of limiting problems like concrete cracking from earthquakes and aging.[3]

Despite the fact that engineers see in the human spine ultimate engineering, a chorus of evolutionists insist that it is a bad design, and attribute this to the evolutionary process having insufficient time to adapt the spine of a knuckle-walking quadruped into the spine of the bipedal human. Nathan Lents, for example, asserts that some lower back problems "are caused directly by design flaws." He continues:

> All vertebrates have disks of cartilage that lubricate the joints be-tween the vertebrae in the spinal column. These disks are solid but compressible to absorb shock and strain. They have the consistency of firm rubber and allow the spine to be flexible while remaining strong. In humans, though, these disks can "slip" because they are not inserted in a way that makes sense given our species' upright posture.
>
> … In humans, the vertebral disks are in an arrangement that is optimal for knuckle-draggers, not upright walkers. They still do a decent job of lubricating and supporting the spine, but they are much more prone to being pushed out of position than the vertebral disks of other animals…. Our ancestors began walking upright about six million years ago. It was one of the first physical changes as they diverged from other apes. It's disappointing, although not altogether surprising, that human anatomy has not had time to catch up and complete this adaptation.[4]

Again, I agree with Lents that evolutionary theory anticipates a poorly designed human spine—a cobbled-together variation on a spine meant for a quadruped. The problem for Lents, as we shall see, is that the scientific research reveals that the human spine is an outstanding design, far beyond anything human engineers have produced.

Optimized for Strength and Flexibility

The vertebral column has the difficult assignment of being both strong and flexible. As with the human foot, these are competing require-ments, so some ingenious design innovations are required. This is exactly what we see in the human vertebral column. It has an optimal structure for a combination of stiffness, strength, and flexibility.

Figure 5.1 shows the vertebrae of the human spinal column. It is an immensely complex engineering structure containing hundreds of mechanical parts. These include around 220 ligaments, 120 muscles, thirty-three vertebrae, thirty-one pairs of nerve branches, and the spinal cord.

One source of flexibility comes through the joints between adjacent vertebrae. Each vertebra fits on top of the other like stackable cups. This stacking allows small movements at each joint. The individual movements are small, but they can add up to big movements across so many joints. The joints allow for bending forward, sideways, and a little backwards. The joints also allow for twisting. The vertebrae and intervertebral discs are tied together with tendons that provide a self-centering mechanism in the column to stabilize it.

A second source of flexibility comes through the S-shape of the spine assembly, as shown in Figure 5.1. The S-shape gives a springiness in the vertical direction.[5] If the spine were straight it would lead to high shock loads for activities like landing from a high jump, because the weight of the body would go straight down the spine with no shock absorption. This S-shape is unique among mammals, which is not surprising because humans are designed to walk on two legs (bipedal) and are the only fully (i.e., obligate) bipedal mammals. Apes have a C-shaped spine, suited to quadruped walking.

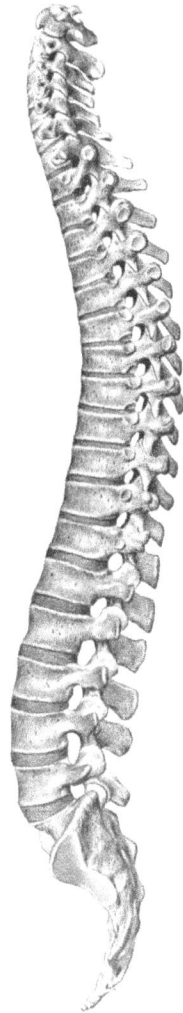

Figure 5.1.
The human spinal column.

Besides having flexibility in all directions, the back is also very strong due to forming a tensegrity structure. Such a structure maintains stability by balancing tensile forces and compressive forces through tension members and compression members. In the case of the human back, the bones take the compressive loads, and the muscles and ligaments take the tension loads.

In engineering it is well known that tensegrity structures are very efficient because they are dominated by compressive and tension loads, which is always more efficient than structures with bending loads. So in this regard the spine has the most efficient type of structure for its purposes. And actually, the human spine is more sophisticated than human-engineered tensegrity systems, because the human back is a smart system where the muscles adjust their tension to adapt to loads. When lifting heavy loads, the 120 muscles increase in tension in order to create a stiffer and stronger structure.

Besides fulfilling structural functions, the vertebral column also protects the delicate spinal cord and provides exit holes so that the nerve roots are spaced equally. These nerve roots, I should note in passing, generally are not strained when the back moves due to there being only small movements at each vertebra.

Spine-Tingling Capabilities

A healthy human spinal column is remarkably flexible, as we see in gymnastic performances. The spine is also remarkably strong, able to carry heavy loads. This is thanks in part to it moving into a smooth curve when flexed, allowing each vertebra to be stressed equally. Some athletes can lift more than twice their body weight above their heads. The world record deadlift is over half a ton.

The back helps protect the spinal cord and organs. In sports like American football, players collide at a combined speed of up to 40 miles per hour. This is surely not what the human spine was optimized for, so it's no surprise that football players sometimes injure it. Yet in the great majority of football collisions, the back provides sufficient stability and protection to allow the players to suit up game after game.

In a lifetime of eighty years, the back performs millions of movements. If a person lives a healthy lifestyle and there is no disease or injury, the back typically works smoothly right into old age. This is more evident in technologically less-developed societies, where individuals often maintain a healthy weight and a good level of fitness from their work.

Self-Assembly and Integration

Of all the miracles of embryological development, formation of the vertebral column and spinal cord may be the most impressive. The nervous system is among the earliest systems to begin forming and the last to be completed. The long period is required because the neural systems are the most complex and most integrated in the body.

When a human embryo is three weeks old and just a quarter of an inch long, a tiny spine begins forming. At this stage, the ribs also can be seen emerging. The early central nervous system begins as a simple neural plate that folds to form a groove. This then turns into a tube, initially open at each end.

The hundreds of intricate mechanical parts of the vertebral column then gradually self-assemble as the embryo grows a perfect spinal column. The millions of nerve pathways also self-assemble in the spinal cord. When we consider that the spinal cord must develop in perfect synchronization with the vertebrae, it becomes evident that we are witnessing a master class in engineering.

One reason engineers are in awe of the vertebral column is that the system is integrated without any fasteners like nuts, bolts, and screws. When you look at an electrical wiring system, it is full of connecting parts that make the system bulky and inefficient. The human nervous system, in contrast, is seamlessly integrated.

Irreducible Complexity, Inexplicable Variety

The vertebral column is a striking example of irreducible complexity (IC). In particular, the integration of the spinal cord and nerve roots with the vertebral discs represents a major IC challenge for evolutionary theory. For example, which evolved first, the holes for the nerve roots or the nerve roots? Neither is useful without the other. And if the column evolved one disc at a time, how did the nerve routes get integrated?

The differences among the spinal columns of different animals also present a problem for evolutionary theory. For example, the sloth has ten neck vertebrae whereas almost all other mammals have seven. The sloth's extra vertebrae allow it to swivel its neck almost all the way around, which means it does not have to move its body so much when looking around—extremely useful for an extremely slow-moving creature that depends for its survival on blending in and being overlooked by predators. But how such a feature could have gradually evolved by mindless evolutionary forces remains a topic of speculation.

Giraffe spines are also unique. The design of their long spines allows the giraffes to reach down to the ground while keeping their heads up. It's not just the length. The cervical vertebrae in the giraffe's neck contain ball-and-socket joints for extra movement. The first and second thoracic vertebrae also contain ball-and-socket joints. Additionally, the joint between the neck and skull permits the giraffe to extend its head almost completely perpendicular to the ground. These unique features are quite different from those of other vertebral columns.

Ironically, the giraffe's long neck is celebrated as something easily explained by evolutionary adaptation, as if the neck merely needed to get a bit longer every few generations, with natural selection rewarding the ability to reach higher and higher tree leaves until nature had evolved the giraffe's long neck. In reality, the giraffe's neck isn't just a lot longer than that of other hoofed mammals; it also involves multiple, complex engineering innovations, which together serve as a particularly dramatic example of irreducible complexity.

Another unique spinal column belongs to the hero shrew, native to the Congo Basin of Africa. This shrew has a super-strong, corrugated, interlocking vertebral column, markedly different from that of other creatures. It has eleven lumbar vertebrae, in contrast to a typical mammal's five. Also, the shrew's ribs are much thicker than those of similarly sized mammals. The spinal muscles are also significantly different. These unique features mean the animal's back can withstand powerful forces, which researchers theorize may serve the purpose of allowing it to pry its way to food sources that it otherwise couldn't reach.[6] The hero shrew is quite an enigma for evolutionary theory but is readily explained from within an ID paradigm.

Someone might ask whether the human spinal column would benefit from the design features found in the shrew, which produce high strength. The answer is that such features would prevent the human back from having the required flexibility. The human spinal column has just the right balance of flexibility and strength for human living.

Functions of the Coccyx

The coccyx is the final section of the spine. It typically consists of three to five fused vertebrae below the sacrum, the bone at the base of the spine. It is a clever design solution for sitting, with the coccyx and the ischium bones functioning as a tripod that helps bear the body's weight and provides balance when one is seated. As with the human foot, a tripod arrangement is the most stable way to make contact with a flat surface.

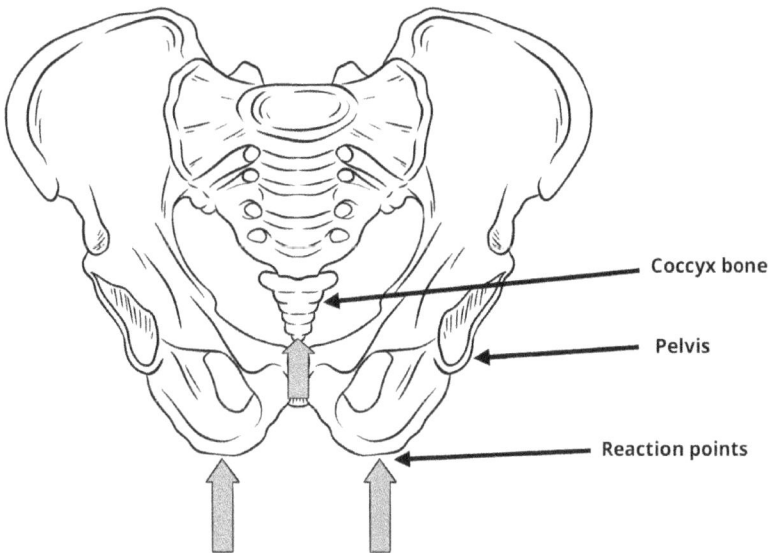

Figure 5.2. Three (tripod) reaction points in sitting.

The coccyx also anchors many tendons, ligaments, and muscles. Some of the muscles form the bowl-shaped pelvic floor, which supports the bowel, bladder, and (in women) uterus. Several of the muscles aid bowel and bladder function. This includes the ability to delay urination and defecation. The coccyx also helps support the spinal cord and aids women in childbirth. The coccyx in females is less curved, affording more room for the baby's head as it passes through the pelvis.

An Evolutionary Tall Tale: The Coccyx as Vestigial Tail Bone

The coccyx is optimized for the functions of sitting, muscle attachment, and childbirth. But despite this, Lents asserts that the human coccyx "is vestigial—a remnant from our ancestors who had tails,"[7] a view that fellow evolutionist Abby Hafer shared.[8] For humans, this supposed evolutionary leftover is, according to Lents, a "pointless cluster of bones" that can be surgically removed with "no long-term complications."[9]

Lents is mistaken here. There can be a loss of strength and function in the pelvic floor (such as the loss of bowel control) if the coccyx is removed, especially if muscles cannot be reattached elsewhere. To take a mundane instance, sitting is often less comfortable without a coccyx.

The fact that a person can survive without a coccyx hardly demonstrates that it is useless, much less that it is a vestigial structure leftover from a tail-swinging tree dweller. There are many parts of the body that can be removed without fatal consequences, such a kidney, eye, or appendix. But there is always a loss of function when such parts are removed. The same is true of the coccyx.

Interestingly, evolutionist Alice Roberts acknowledges that the coccyx is important for sitting and for muscle attachment.[10] However, she still sees the smaller bones that make up the coccyx as having evolved from an ancestral tail. Such a conclusion is weakened by the fact that the coccyx is optimal for human function.

In engineering it is common to see a structural concept used for different applications. For example, structural columns are used not only in buildings but also in bridges and cars. While there is a clear family resemblance among these columns—enough that we call all of them *columns*—there also are distinct differences among them due to differing functional (and aesthetic) considerations. And here, obviously, there is no question of one type of column blindly evolving from another. They are all products of intelligent design. So the fact that a column of bones is used both in humans and in animals with tails is not surprising from a design point of view. And indeed, that perspective would lead us to expect a clever designer to reuse, with modification, various fruitful design concepts.

A quick point regarding terminology: The coccyx is colloquially referred to as the tailbone. This is not inaccurate in that the coccyx is found at the tail end of the spine. However, because of evolutionary teaching, people mistakenly assume that it was dubbed a tailbone because evolutionary theory teaches that it evolved from a tail. In fact, the term in English predates the rise of evolutionary theory in England and Europe by hundreds of years.

Whatever we call the coccyx, it is important to recognize that it is an optimal design for humans and shows no sign of being an evolutionary leftover.

Answering Claims of Bad Design

As noted above, Lents claims the human spinal column is a flawed design because the vertebral disks are not inserted in the right way for upright stature. Lents makes this claim not because of scientific evidence but because this is what evolutionary theory anticipates. Lents assumes that there was not enough time for us to fully evolve from a knuckle-walking quadruped to a biped and, based on this evolutionary thinking, he's on hair-trigger alert to see bad design in the human back. This, along with his lack of engineering knowledge, leads him to overlook the fact that the vertebral column makes excellent engineering sense. Engineers who understand mechanics view the human back as an optimal design and are keen to copy it, as in the case of the well-funded efforts to build better bridges, noted at the beginning of the chapter.

Alice Roberts, while a bit more sophisticated in her analysis, also has been drawn by her evolutionary paradigm into mistakenly regarding the human spine as a poor design. "Human spines are 'good enough' as far as evolution is concerned," she writes, "but plagued with problems, especially in later life."[11]

I agree with Roberts that if the human spine had evolved by Darwinian means, it would be only just "good enough" and not anything we would take for ingenious design. But the reality is that the human spine is manifestly a masterpiece of engineering, far beyond what we could expect from a just-good-enough-to-survive evolutionary process. Like Lents, Roberts makes the mistake of not differentiating between problems caused by poor design and problems caused by other factors, such as those arising from misuse, disease, or the aging of any engineered system.

Yes, lower back pain is all too common in modern life, but this is due not to poor design but to several other factors, including poor lifting technique, poor posture, abuse from certain sports, excessive body weight, and poor fitness. A combination of two or more of these factors further increases the risk of back problems. For example, when a person is overweight *and* has weak abdominal muscles, this puts extra strain on the back.

The modern sedentary lifestyle (for which humans were not designed!) is bad for the back because of too much sitting and a lack of exercise. Join this to the weekend-warrior syndrome (sedentary weekdays punctuated by aggressive sports), and the result is a particularly potent formula for back injury.

In modern life it is helpful to have routines that include periods of standing and exercise. Simple changes like occasionally using a desk at standing height can be very good for back health. For those who have to do a lot of lifting or bending in their work, it is especially important to have good posture and lifting technique while also maintaining good core strength. It is also important to know your limits. In some areas of life, special care is needed to avoid problems, and the back is one of those areas.

A Stack of Evidence

The human vertebral column is recognized by the biomechanics community as ultimate engineering. There are some similarities between the human back and the chassis of a Formula 1 race car. Both have to be very strong and robust and yet have some flexibility for shock absorption. The difference is that there is far more precision engineering and functionality in the human back than in even the best Formula 1 chassis.

We have a choice about how to approach this subject and how we explain it to others. We can face and frankly describe how scientific research has shown the spine to be masterfully optimized for the human body, or we can repeat the inaccurate and uncritical evolutionary just-so story about a poorly designed back jury-rigged from the back of a tree-dwelling knuckle-walker. There is an urgent need to move beyond discredited just-so stories and communicate the perspective backed by science.

6. Mouth, Nose, and Throat

Two Views of the Mouth, Nose, and Throat

"Organisms," comment the authors of a 2019 article in the journal *Integrative and Comparative Biology*, are "phenomenal multitaskers" due to "their ability to use the same structures for multiple functions."[1] This capacity for multitasking is of increasing interest to biologists and bioengineers. Also in 2019, the Society for Integrative and Comparative Biology dedicated a symposium at their annual conference to "Multifunctional Structures and Multistructural Functions: Functional Coupling and Integration in the Evolution of Biomechanical Systems."[2] Although the symposium title dutifully name-checks evolutionary theory, and several of the speakers strove to make sense of multi-functionality in light of evolutionary theory, much of the symposium involved a straightforward engineering analysis of multifunctioning biological systems in the here and now, a form of analysis that actually poses a big challenge to evolutionary theory.

Five years after this symposium, I wrote the review paper mentioned earlier (Chapter 3) explaining how multifunctioning is a key reason for the exceptionally high performance of biological systems, such as the human wrist and knee.[3] The paper explains that multifunctioning involves trade-offs but always with a net benefit in biological systems. I urged that if engineers are to improve the performance of robots, they need to learn how to produce multifunctioning systems like those found in biology.

One of the multifunctioning systems I gave to support my case was the human throat, an ingenious multifunctioning design that has astounded scientists. And yet, because of her evolutionary paradigm, Abby Hafer insisted that the throat is an example of bad design. This chapter will show why multifunctioning in the throat is not a design weakness but a brilliant design feature.

The Swallowing Process

It is easy to take swallowing for granted, but a closer look reveals an incredible feat of biomechanics involving the precision coordination of over twenty muscles, including those in the mouth, pharynx, and upper esophagus. Figure 6.1, a cross-section of the mouth and throat, shows the main regions. A key part, the epiglottis, forms a flap that closes the airway during eating. The hyoid bone, in front of the epiglottis, allows a wide range of tongue, pharyngeal, and laryngeal movements by bracing these structures. The hyoid bone is special because it is a "floating bone"—not connected to any other bone. This means the muscular control of the hyoid bone must be very sophisticated and precise.

As is true of many parts of the body, the throat is a sophisticated reconfigurable system. The main reconfiguration occurs when the epiglottis shuts off the air passage to the lungs, converting the throat from an air-breathing system to a food-eating system.

The teeth have an obvious role in initiating the digestive process. But the tongue is also very important in eating and swallowing because it manipulates and moves food in the mouth, and cleans the mouth. The tongue has four intrinsic muscles (within the tongue) and four extrinsic muscles. The four intrinsic muscles change the shape of the tongue while the four extrinsic muscles move the tongue around. After the mouth's saliva and chewing action begin the digestion process (Stage 1), the steps for swallowing are as follows:

Stage 2 (Oral transport)

With the mouth closed, the soft palate rises to close off the passageway between the nasal and oral cavities. The tongue rolls backward, propelling food into the oral pharynx. These actions can occur automatically without the need for us to focus our attention on them.

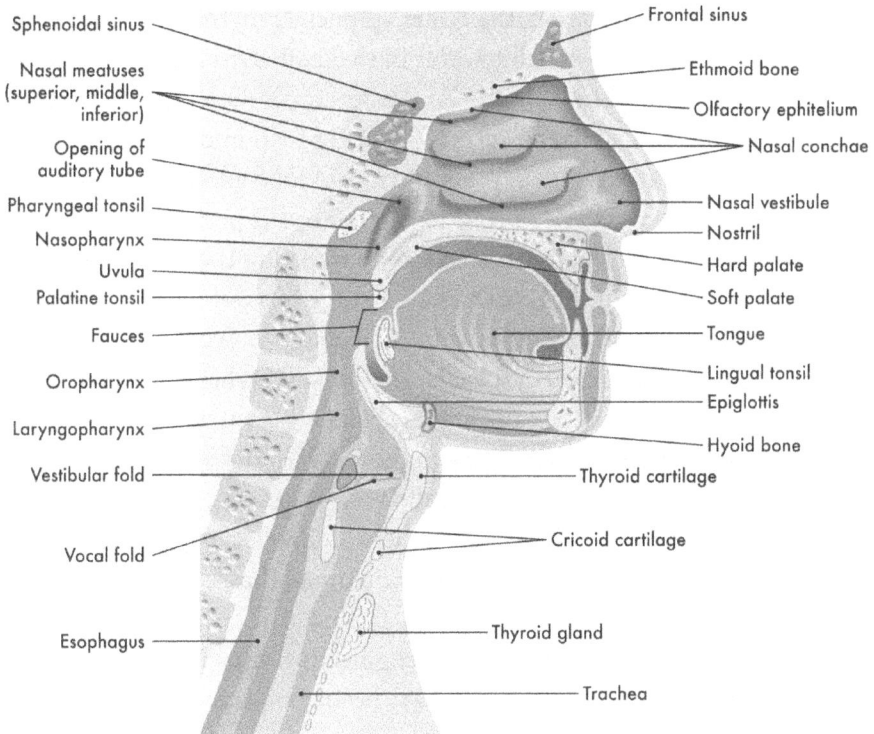

Figure 6.1. Anatomy of the human mouth, throat, and nose.

Stage 3 (Pharyngeal)

Breathing is paused as the larynx rises, closing the glottis, the opening to the air passage. This raises the Adam's apple. Pressure in the mouth and pharynx drives food toward the esophagus. A muscular constrictor (the upper esophageal sphincter) at the top of the esophagus relaxes to open as food approaches.

Stage 4 (Esophageal)

Food moves from the pharynx to the esophagus, and then the upper esophageal sphincter immediately closes, preventing food from returning into the mouth. Breathing resumes after the larynx lowers and the glottis reopens. From when food leaves the mouth until it passes the upper sphincter, only about one second elapses, during which all these precise mechanisms occur.

After the food passes the upper sphincter, rhythmic muscular contractions create peristaltic waves in the esophagus that usher the food down toward the stomach. At the bottom of the esophagus, the lower esophageal sphincter relaxes, allowing food into the stomach. Then the sphincter closes to prevent reflux of food and gastric juices.

Swallowing properly is vital for survival, as it performs two essential functions of ingesting food and protecting the airway. That a properly functioning throat is essential for survival under normal circumstances is underscored by this: An inability to swallow correctly, known as oropharyngeal dysphagia, leads to serious health complications, including aspiration pneumonia, which can prove deadly.

Fortunately, the system works in the vast majority of cases. Babies are able to perform the complex swallowing process immediately after birth as they breastfeed. Adults swallow around a thousand times a day and many millions of times in a lifetime. That such a complex system routinely functions, and lasts as long as it does, suggests an extraordinary level of sophisticated engineering. This is underscored by the fact that human engineers have been unable to create an artificial throat replacement.

The Human Voice: A Masterpiece of Engineering

The throat's ability to facilitate both eating and breathing is a masterpiece of design shared by animals generally, but some unique design properties of the human throat allow humans to excel at vocalization.

The human voice is extremely versatile, able to produce a wide range of sounds with great clarity and control. There are four main elements of voice production. There is the air flow from the lungs that powers the voice. There are the vocal cords themselves, which function as a vibratory instrument. There are resonating cavities of the throat and mouth, which magnify the sound. There are the tongue, lips, jaw, and palate, which change the pitch and articulate the sound.

A key requirement of speech is an agile tongue that can form precise shapes so as to change the layout of the vocal tract to control the pitch of the sound. As we speak, the shape of our tongue changes constantly in order to produce different sounds. The lips also play an important role in speech. For instance, closing the lips makes it possible to build up pressure and suddenly release pressure to make the p and b sounds.

When singing, the human voice is able to produce great variations of pitch and tone. Many regard the human voice as far superior to any wind instrument, such as a flute or oboe. Minoru Hiran, a world-leading expert in the area of laryngology, explains:

> Singers produce great varieties of voice in pitch, loudness, and tonal quality with the use of only one sound generator; that is, a pair of vocal folds. Any singer, however great, has never had more than a pair of vocal folds. This is in surprising contrast to many musical instruments that require multiple sound generators in order to produce a variety of tones…. Comprehensive studies of the anatomy and physiology of the vocal folds and the muscles that control them reveal that humans exhibit remarkably complex control over the characteristics of their vocal folds. Alterations in the length, stiffness, shape, and other characteristics combine with changes in air flow, resonance, and other activities to permit such extraordinary diversity.[4]

Incredibly, around one hundred muscles are used during singing, most of which are not under our conscious control. We often take speech and singing for granted, but they are both highly complex processes. A rule of thumb for child development is "walking at one, talking at two." Learning to walk is extremely complex, and yet it takes twice as long to learn to talk. This shows the complexity of speech. While skilled engineers have produced clever artificial electrolarynxes for people who have lost their voice, the designs are very crude compared to the real thing.

The Sense of Smell

Like the mouth, the nose is multifunctioning, having several important functions including air intake, air filtering, air conditioning, and the sense of smell. The sense of smell (olfaction) is the ability to sense volatile chemicals suspended in the air.

We are sometimes told that humans have an inferior sense of smell to dogs because dogs can detect tiny amounts of odor. However, this is not the whole story. Humans excel at differentiating among different types of smell. The typical person can distinguish up to 10,000 distinct odors. Examples of odors we can detect are the sweet aroma of esters produced by fruit, the sweet scent of flowers, and the pungent smells of rotting food. Odorant molecules pass through the nostrils on their way to the nasal cavity, dissolving in what is known as the olfactory epithelium. Embedded in the olfactory epithelium are odorant receptor proteins that detect smell. The human olfactory system has up to twenty million odorant receptors.

It is estimated there are about a thousand different genes that code for roughly four hundred different odorant receptor proteins. The odorant receptor proteins are complex structures with molecular configurations that are currently poorly understood. These receptors, and our sense of smell generally, have impressed researchers. One paper describes the human sense of smell as possessing "exquisite complexity."[5] Another paper, in the journal *Nature*, describes odorant receptors as "engineered" to provide excellent discrimination of smell.[6] This makes good design sense because it is more useful to humans for us to be able to differentiate between many types of smell than, as with dogs, to be able to detect extremely faint odors.

Multifunctional: A Strength, Not a Weakness

Despite the abundant evidence of the ingenious design of the throat, Abby Hafer insisted it is badly designed, increasing the choking hazard due to the same structures being used for eating and breathing. "A better-designed system would keep the tubes for air and food separate, to avoid unnecessary fatalities," she writes. "If we were designed, why did the Designer do this job so badly?"[7]

Hafer was making the mistake of following an evolutionary paradigm rather than the scientific evidence. Multifunctioning in the throat is not a design weakness; it's a design strength. The field of biomimetics is seeking to imitate the extraordinary multifunctionality found in countless biological systems, because such an approach offers so many design benefits.

In engineering it has long been known that multifunctioning is a way to maximize performance. A good example is the smartphone.

One could imagine a pocket-sized device that could do any one thing that a smartphone does, and do it better than the smartphone. But a smartphone packs many functions into a single pocket-sized device, and performs many of these functions more than adequately. Yes, one could, in principle, have a separate device for each function performed by a smartphone, but carrying all those devices around would be cumbersome, as would switching from one function to the other. And some activities would become well-nigh impossible. Imagine a physician going for a long run while using a GPS device to navigate, another device to listen to podcasts, another device to record his heart rate patterns during the run, another device to take emergency calls, another device to take notes during and after an emergency call, and another device to access medical information on the internet in case of an emergency call. In theory he could have six separate devices that performed each of these six functions slightly better than his single smartphone, but in practical terms, the whole endeavor would be rendered slow, cumbersome, and borderline unworkable.

The human nose/mouth/throat complex is like the above example. It handles eating, breathing, and nuanced verbal communication along with tasting and smelling. Other animals, including other primates, have throats that combine eating and breathing. But when we compare the human throat to that of other primates, we find some subtle and not so subtle differences. For instance, the larynx (voice box) in humans is situated lower in the throat. This might (emphasis on the word *might*[8]) increase the likelihood of choking, but it also allows for a larger pharyngeal cavity, which in turn enhances our ability to make a wide range of sounds, crucial to our ability to master highly complex spoken language. This is just one example of how, when assessing the design of the multifunctional human throat, the downsides—such as a need to be a little careful when eating—must be fairly weighed against the sizable advantages in the tradeoff.

To take another example, multifunctioning avoids duplication of parts. The tongue, teeth, palate, jaws, and cheeks are used for both eating and speaking. To duplicate all these together with the associated muscles would be inefficient and cumbersome. If you just consider the tongue alone, which has eight muscles, it would be very inefficient to produce two tongues. One could imagine a humanoid where the eating and breathing functions of this system were handled by separate organs. But then the humanoid's head and neck would have to be much bigger, making the individual less agile and complicating the birthing process.

Another advantage of the throat's elegant combination of functions is that it provides two airways. The nostrils are used when there is a need for breathing moderate quantities of warmed, humidified, filtered air. But then the mouth allows rapid entry of much larger quantities of air when, for example, you need to run or swim hard. Integrating the nasal passageway with the mouth also aids in the tasting of food, because smell and taste occur in the same set of passageways, and the sense of smell is known to enhance the tasting experience.

The human throat and the smartphone both illustrate what is widely acknowledged in engineering: The strategy of combining multiple functions into a single organ or device requires considerable skill and, by affording greater compactness, can lead to a host of performance benefits. We saw this principle at work in the chapters on the wrist and ankle joints.

Yes, the fact that the throat combines the functions of eating, early digestion, and breathing means care must be taken not to gulp air when eating. However, there are many things in life where care is needed. The eyes are a delicate part of the body where care is needed to protect them from dirt. Our fingers are delicate and can be injured. If we had claws instead of fingers, they would be less vulnerable to scrapes and burns and would be better for scratching and ripping certain things apart. But the price would be a greatly diminished sense of touch and less dexterity. Biological systems, like any engineered systems, involve trade-offs. The fact that swallowing requires a bit of care is the reasonable price for the considerable benefits of our multifunctioning throats.

When people get old, swallowing can get difficult, but many things get difficult when we're old. Such problems are not a reflection of bad design but an example of an issue common to all engineered systems, however well designed—degradation over time.

It is important to note that evolutionists who criticize the design of the throat do not produce a detailed solution for a credible alternative design. For example, Hafer claimed that humans would be better off with a blowhole like whales have: "The whale's respiratory system is completely separate to the digestive system... If the Creator could do that for whales, I don't know why he couldn't do it for us."[9] But Hafer never offered a detailed blueprint for an improved alternative design.

It would be interesting to see one—or better yet, a realistic animation of a humanoid with such a blowhole design. I won't hold my breath. What works for a multi-ton whale would have unacceptable downsides for a creature our size. Such a design would, as noted, be prohibitively bulky. It also would trade away the advantage of connecting the sense of smell with the sense of taste. And just at a purely aesthetic level, it would be a very inelegant design for humans.

Purposeful Overdesign Supports Intelligent Design

Like the human hand, the human voice is an example of purposeful overdesign. The intelligent design framework readily accommodates such an occurrence, while such overdesign is an extremely awkward fit with evolutionary theory. The evolutionary paradigm anticipates a human voice that has evolved only just enough to accomplish gene survival. On this point, the grunts, hoots, and shouts of other primates may not be particularly daunting to evolutionary theory. It's not hard to imagine natural selection awarding a series of evolutionary baby steps on the way to such capacities. But the human voice, with its capacities so far beyond what is required to get by? This poses a qualitatively different challenge to evolutionary theory.

One could make recourse to Darwin's patch theory, sexual selection, with our extraordinary vocal capacities construed as evolving to woo sexual partners. But where is the wonder-working force of sexual selection in the vocal capacities of all the other mammals? Why didn't any of them get anything approaching our nuanced vocal abilities? And if we shift our focus from natural selection to the mutational side of the mutation/selection mechanism, the question arises, what is the plausible, detailed mutational pathway for gradually evolving the many intricate and ingenious engineering solutions required for the human voice? The evolutionist has no good answers for such questions—little more than a shrug and a call to trust the evolutionary paradigm.

One might also appeal to the idea of neutral evolution. Without going into excessive detail, we can say that the theory of neutral evolution looks for ways that a genome can experiment with mutations unhindered by that harsh taskmaster, natural selection. When might such selection-free evolution occur? Various scenarios are proffered. These include genetic drift, which occurs in small populations; mutations in regions of the DNA that do not code for proteins; and relaxed selection in undemanding environments. Then, when some new complex feature emerges in the neutral evolution sandbox, it can escape the sandbox with a bit of luck, as for instance when the environment changes and a previously neutral feature suddenly proves wonderfully useful in the new environment.

The imaginative storytelling is first rate, but the devil is in the details—or rather, in the lack of details. Neutral evolution proposes to set natural selection aside, at least temporarily. Now the evolutionary process of random genetic mutations is said to blindly and purposelessly construct this or that intricate system or organ, one that requires numerous parts precisely fitted and arranged before it has any useful function. Natural selection would stop the evolutionary process in its tracks, since the evolving feature is dysfunctional till the end point. So, sideline natural selection for the moment. Problem solved, right?

No, because natural selection was Darwin's master stroke, brought in to tame the astronomically prohibitive odds against a series of random variations successfully building sophisticated biological forms in the history of life. Without something like natural selection, we are back to the old problem of monkeys banging away on typewriters and the zookeepers hoping the monkeys will produce *Hamlet*.[10] Probability calculations tell us that the industrious monkeys won't luck onto so much as a quatrain from one of the bard's sonnets, not if all the world were a stage full of typing monkeys banging away till Doomsday.

Evolutionists, then, face a pick-your-poison dilemma. They can't live with natural selection, since it stops gradual evolutionary progress in its tracks in cases where several mutations are needed together to produce anything useful. But they can't live without natural selection, either, for the reason described above, the very reason Darwin appealed to it in the first place.

So, why don't evolutionists give up on neutral evolution? Some have, but my sense is that for many evolutionists, it's just too valuable a tool in their storytelling toolkit. It may break down under unflinching cross-examination, but that's easily enough avoided in a scientific community where evolutionary theory is the dominant paradigm. Indeed, so insulated are most evolutionists from exposure to sustained and thoughtful critique of the theory that they would need to go out of their way to encounter it, which is easier said than done when one hardly has enough time as it is for a busy teaching and research schedule.

The Human Voice Is Telling Us Something

The human voice is a prime example of ultimate engineering, and of engineering vastly beyond that required for survival. This is entirely consistent with an intelligent design paradigm. And if we move beyond the strict confines of intelligent design reasoning and avail ourselves of theological considerations, the evidence for design in this case becomes even clearer. This is because if the designer has made humans to be the stewards of the earth and, among all the animals, to be in a uniquely intimate relationship with our maker, as is taught in the Judeo-Christian tradition, then we can expect that humans would have great powers of communication.

And again, at a purely design level, we should be impressed that this speaking/singing instrument can be reconfigured to eat as well as enhance breathing during physically demanding activities. All three functions are masterpieces of design separately, and are exponentially more so in being combined into a single compact instrument.

The issue is a good example of how biologists can benefit from engaging with engineers and engineering principles to better understand and properly appreciate the design of biological systems—principles such as multi-objective optimization and design for compactness.

I submit to you that when we avail ourselves of this perspective, and do so with an open mind, the ultimate engineering we see in this eating-breathing-tasting-smelling-speaking-singing apparatus speaks loudly of a very intelligent design.

7. The Jaw

Two Views of the Human Jaw

A systematic review of human chewing simulators was reported in a 2020 issue of the *Journal of Clinical and Experimental Dentistry*. The researchers studied the best robots from around the world and found that none could match the sophisticated motions of the human jaw. "No chewing simulator offers all the characteristics necessary to reproduce human masticatory movements and forces under the humidity and pH conditions of the oral cavity," Sergio Soriano-Valero and his colleagues reported.[1]

The reason engineers are trying to recreate the human jaw motion is that there is always a need to test dental materials and structures in the lab in conditions that best represent actual human eating. I myself was involved in designing an advanced robotic chewing simulator with a team at Bristol University. One of the papers we published was cited in that 2020 review.[2] Our work was successful enough that we were invited to present our chewing robot at a prestigious science exhibition in London at the Royal Society in 2009.[3] But we weren't in any danger of getting a big head, since our design—like all the other chewing simulators out there—falls well short of the human original. From our work on the jaw simulator, we learned just how masterful the design of the human jaw is.

The jaw muscles are incredibly strong. The second molars can exert over 225 pounds of bite force (1,000 newtons). And teeth have an extreme hardness, with the ability to chew millions of times throughout an adult lifetime.

Researchers have given high praise to the efficiency of the human jaw, with one article from a leading journal concluding, "Our findings show that the human masticatory apparatus is highly efficient, capable of producing a relatively powerful bite using low muscle forces."[4]

Despite such findings, evolutionists such as Abby Hafer have asserted that parts of the jaw are poorly designed. This chapter will show why that claim is misguided and how the human jaw is optimal for human use.

An Ingenious Hinge Joint

The jaw has an ingenious hinge joint, allowing the lower jaw to slide as well as hinge up and down. Such a joint has a wonderful technical name as complex as the joint itself: a ginglymoarthrodial joint. The individual joints connecting the jawbone to the skull are the temporomandibular joints (TMJ), so called because the lower jaw is the mandible and the part of the skull where the lower jaw attaches is the temporal bone. (The upper jaw is the maxilla.) The mandible has two protrusions called condyles, and these articulate with two concave notches in the temporal bone.

The TMJs (one on either side of the jaw) enable three types of movement. The TMJs can (1) open and close the mouth (elevation and depression), (2) move the lower jaw out and in (protrusion and retraction), and (3) move the lower jaw from side to side (lateral motion). These combine to give the jaw its complex up-and-down and side-to-side movements during chewing. Without this versatility our jaws would be much less effective at chewing and preparing many kinds of food for further digestion.

Four main muscles produce the lower jaw's complex movements. These are the masseter, medial pterygoid, temporalis, and lateral pterygoid muscles. The masseter muscle closes and protrudes the jaw and is situated in the cheek area. The medial pterygoid muscle closes and protrudes the lower jaw but also moves the lower jaw sideways. The temporalis muscle is involved in closing and retracting the jaw and is situated on the temporal bone in the skull. The lateral pterygoid muscle mainly serves to pull the heads of the condyles out of the

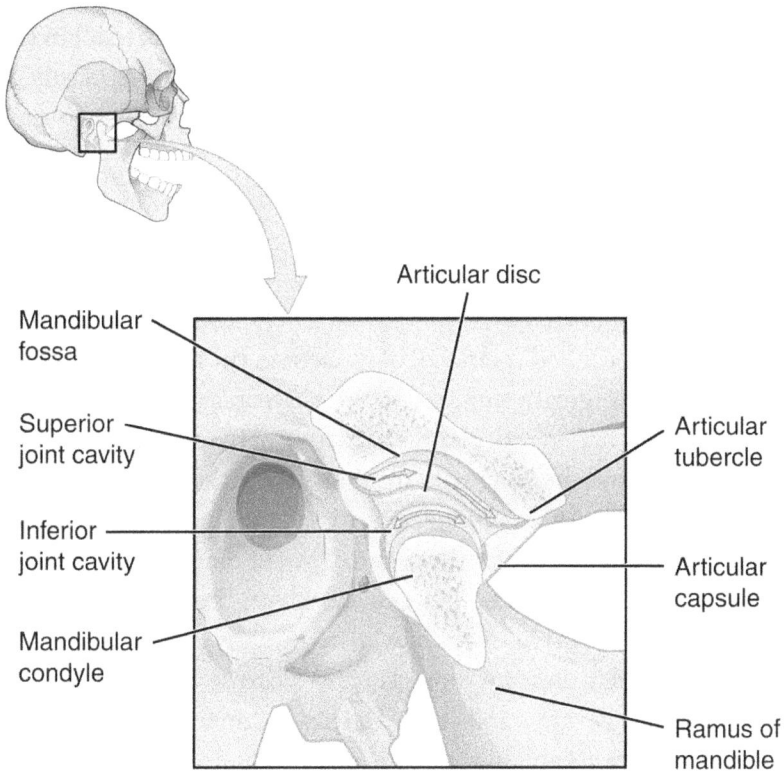

Figure 7.1. The human jaw, with a closeup of the temporomandibular joint.

mandibular notches to protrude the lower jaw. The lateral pterygoid can also open the jaw.

These four jaw muscles work together to produce complex chewing motions with a powerful force. Engineers have struggled to replicate the motion and forces of the lower jaw. An added challenge is replicating the complicated profile of the teeth. For this reason, dentists are still hoping for improvements in chewing simulators.

The human jaw has some very impressive capabilities. Typical mastication forces range from 70 to 150 newtons, with a maximum force of up to 1000 newtons (225 pounds of force). The maximum closing force of 225 pounds is enough to allow a fit slim person to hang from their teeth—a feat often demonstrated by circus gymnasts. The temporomandibular joints are two of the most frequently moved joints in the body, moving many millions of times in an eighty-year lifetime.

The enamel on the outside of teeth is the hardest material in the human body and harder than steel. Tooth enamel consists mostly of hydroxyapatite, a complex crystal material. Hydroxyapatite crystals are needle shaped and provide great rigidity and hardness.

The Special Articular Discs

The secret to the remarkable flexibility of the jaw joint comes from an ingenious component called an articular disc, shown in Figure 7.1. It allows both linear and rotational movement in the jaw, joint motions that combine to greatly enhance the effectiveness of chewing. Each side of the jaw has an articular disc, situated between the condyles in the lower jaw and the notches in the skull. The articular disc is a thin, oval plate made of fibrous connective tissue that can both move and change shape as needed for the optimal functioning of the joint. The disc divides the TMJ joint into two sub-joints, the lower and upper sub-joints, each with its own synovial cavities.

The articular disc can move laterally because it is flexibly connected to the lower jaw and skull. The disc is connected to the lower jaw with ligaments and to the skull with a tendon. The articular disc allows the lower jaw to slide forwards as well as rotate downwards (i.e., open the mouth). When the jaw is opened only a little (up to about 20mm), the jaw mainly rotates. Beyond 20mm the jaw simultaneously rotates and slides forward.

The unique articular disc, both agile and durable, helps explain why engineers find it so difficult to replicate the human jaw. Engineers do not yet have the materials and technology to replicate the articular disc. In particular, engineers cannot copy the seamless connections between the articular disc and bones, nor can they copy the synovial fluid lubrication system.

The Jaw as Anchor

In addition to facilitating the chewing motion, the jaw also anchors the teeth via roots in the jaw's alveolar bone.

A difficult technical challenge for the jaw in its role as anchor is that two sets of teeth are required, one for early childhood and one for adulthood to accommodate the larger size of the adult jaw. Humans have baby teeth from around age one to the early teens, with baby teeth beginning to be replaced at around age six. After tooth replacement is complete, humans have thirty-two adult teeth. These teeth include eight incisors, four canines, eight premolars, and twelve molars. Orchestrating a system of automatic teeth replacement is a daunting challenge beyond anything our best engineers can manage. The system solves the problem of matching teeth to a steadily growing jaw through childhood. The process is doubly impressive in that the replacement is designed to occur not in short order but over several years of childhood, meaning the child's mouth is never missing very many teeth.

Problems for Evolution

It is difficult for evolutionary theory to explain why there are so many differences between the human jaw and the jaw of apes. The human lower jaw is not only much smaller than the chimpanzee jaw but, as we can see from a side view (Figure 7.2), it also features a very different center of rotation.[5] Viewed from above (Figure 7.3), we see that the ape jaw is nearly rectangular, and the human jaw oval. The orientation of the jaw muscles is also quite different between chimp and human, with the temporalis muscle in the human jaw attaching to a different part of the lower jaw (the coronoid process rather than the zygomatic arch as in other primates).[6]

Chimpanzee   Human

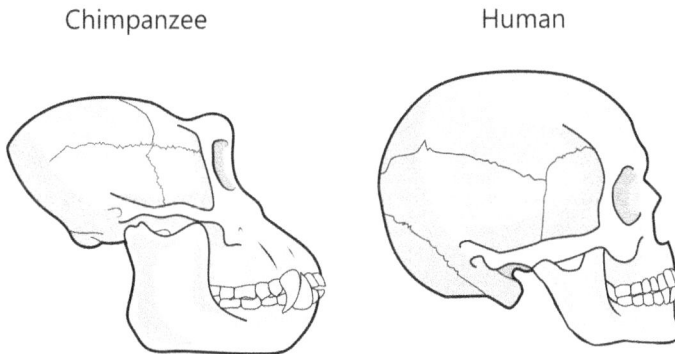

Figure 7.2. Side view of chimpanzee jaw and human jaw.

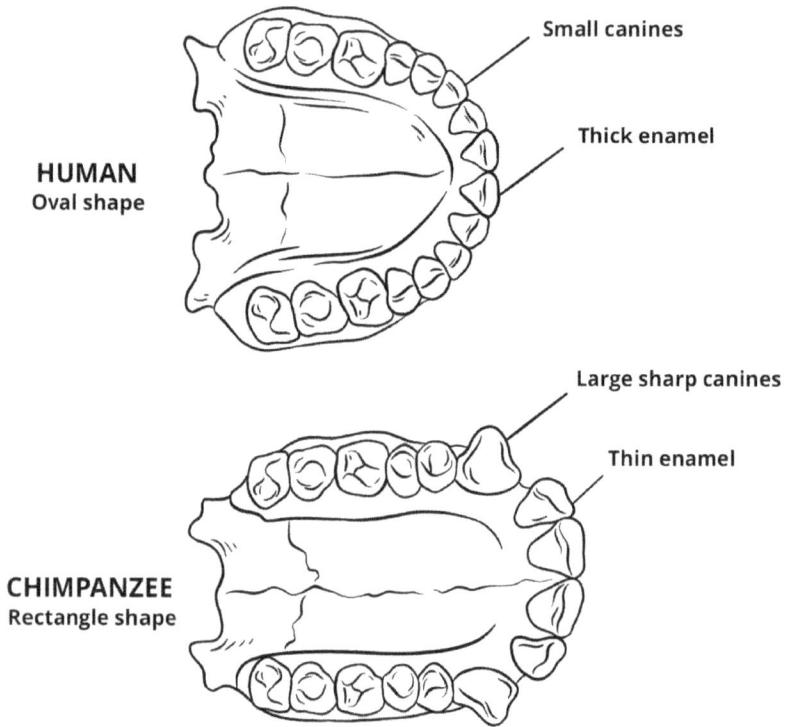

Figure 7.3. Overhead view of human jaw and chimpanzee jaw.

Evolutionary theorists have proposed stories to explain all these differences in Darwinian terms, but as is generally the case, the stories are long on imagination and short on genetic and engineering particulars. Which came first, the ability to speak and then the jaw changes to aid the speech? Or vice versa? And what about a detailed evolutionary pathway where function is maintained or enhanced every mutational step of the way? And what about the origin of animal jaws in the first place? For this there is the usual stringing together of a few fossils in a proposed sequence, with countless steps missing and in their place only what evolutionary theory best excels at generating, promissory notes of future discoveries.

Consider the articular discs. They are not unique to humans, but they are an example of irreducible complexity. The articular disc is a component that only has a use when it has the precise shape along with the correct muscles, ligaments, and tendons to function properly. Without all that, it's just a proverbial pebble in the gears. Evolutionists have struggled to explain its origin. For example, a 2020 paper in the journal *Frontiers in Cell and Developmental Biology* offers the following observation: "Little work has sought to understand the TMJ in an evolutionary and comparative developmental biology context. This is despite the crucial role that the formation of the TMJ has in mammalian evolution. An important part of the TMJ is the disc that cushions its action. The origin of the disc is uncertain."[7]

The authors of the article say that little effort has been made to explain the evolutionary origin of the TMJ despite its importance. Actually, it would be difficult to know how much work has been expended on the question, since a failure to discover a viable evolutionary pathway is unlikely to result in a published paper on the topic. Their mention of "little work" on the subject is, rather, a sign that researchers are struggling to think of how such a complex joint could evolve. The authors go on to admit that the origin of the articular disc is "uncertain," a common euphemism for *we're in the dark on this one.*

Answering Claims of Bad Design

Despite the high praise from engineers focused on the human jaw and accompanying teeth, some evolutionists remain unimpressed. Hafer, for example, said, "Sharks continuously replace their teeth throughout their lives.... Those teeth don't have time to get cavities, so does the Creator like sharks more than humans?"[8]

Before we consider Hafer's proposal of giving humans the shark's system of continuously replacing teeth, let's pause to consider the matter of cavities. It is abundantly clear that just as the human body isn't optimized for a diet heavy in processed sugar, human teeth aren't either. As any dentist will tell you, sugary snacks and drinks promote tooth decay. Archeologists have found that tooth decay was far less common with hunter-gatherer cultures. What this tells us is that the prevalence of tooth decay, if not its outright existence, can be chalked up to misuse of the equipment.

However, the Hafers of the world might still press the point: Wouldn't a properly intelligent designer have foreseen even the modest levels of tooth decay among hunter-gatherers, not to mention the higher rates in agricultural societies, and opted for the shark-teeth solution? It turns out, there are clear reasons why shark teeth are not appropriate for humans. Individual shark teeth are not rooted but held with soft connective tissue. This works for the shark because their mouths are big enough to accommodate large amounts of connective tissue. In the case of human jaws, there is not enough room for this type of design. Because the human jaw is small compared to a shark's, the best design solution is to have deep-rooted teeth embedded in the jaw. This is a compact design optimal for humans. Ignoring this fact is another example of an evolutionist being oblivious of the engineering details of a system and the elementary design principle of trade-offs. (In fairness, there are evolutionists considerably less prone to these sorts of reflexive inferences to poor design, even if much of the popular marketing machinery for evolutionary theory depends on just such sloppy claims of poor design.)

OK, but what about wisdom teeth? Evolutionary theory teaches that humans evolved from ape-like ancestors that had larger jaws and teeth. It is claimed that the process of evolving smaller jaws caused overcrowding and hence problems such as accommodating third molars—that is, wisdom teeth. The third molars are situated at the back of the mouth and are the last to emerge, often leading to dental collisions, pain, and extra business income for dentists. Moreover, while most people have their third molars, a significant portion of the population do not, as if evolution is in the early stages of solving this overcrowding problem.

But there is a modest explanation for our current problem with wisdom teeth that doesn't saddle us with the challenge of finding a macroevolutionary pathway from ape-like jaw to human jaw. Research has shown that problems today with the third molars may be caused by diets that put few challenges on our jaws. Our diet is heavy on highly processed and cooked foods. Long gone are the days when our very capable jaws struggled with uncooked meat, dried meats, and a frequent diet of raw, fibrous vegetables. In past societies, a tougher diet exercised the jaw muscles far more, and this helped the jaws develop more fully, giving the teeth more room. As a consequence, problems with wisdom teeth were relatively rare.[9] In contrast, our modern diet of soft, processed foods is so unchallenging for our jaws that it fails to promote proper jaw development, leading to problems with our wisdom teeth. The phenomenon parallels the rise in back problems in our modern sedentary and often overweight population. The human body wasn't designed for such pampered treatment.

Interestingly, the average tooth size in modern human populations appears smaller than in ancient humans,[10] suggesting a built-in capacity for microevolutionary adaptation to address the shrinking jaw. It may simply be that the soft modern diet has outstripped that helpful capacity to adapt.

This diet, loaded with soft foods and refined carbs, is causing other problems, including a spike in diabetes and other diseases. One could complain that a competent intelligent designer would have made the human body capable of thriving on any sort of diet. But this is an unrealistically demanding standard that we hold no human-engineered system to. The capacity of humans to manage on a wide variety of diets—from the blubber-heavy, vegetable-poor diet of the traditional Inuit society to the Mediterranean diets of the Greek islanders of earlier generations—is testimony to how much dietary diversity has been built into the human species. Insisting that the human body exceed this impressive versatility, to the point of being wholly immune to dietary abuse, is hardly reasonable.

Jaw-Droppingly Well Engineered

The human jaw, along with the mouth, is supremely well designed for eating and speaking. The floating articular disc—to highlight just one ingenious aspect of the jaw—is a masterfully designed component that gives the temporomandibular joints their finely tuned range of movement. It's so masterfully engineered that our best engineers struggle to copy it and, with it, the human jaw.

Things like wisdom teeth problems and cavities at most suggest isolated design weaknesses rather than a lack of high-quality design, much less a lack of intelligent design altogether. But in fact, when we look more closely at these supposed design flaws in the jaw and teeth, we find a more complex picture. There are good engineering reasons for a designer not to have employed the same strategy of regular teeth replacement found in sharks. And the great frequency of cavities and wisdom teeth problems appears to stem from the soft, sugary, processed modern diet. Ideally, our teeth and jaws would be immune to such misuse, but this is an unrealistic standard.

It's like faulting a Ferrari because it had a puncture or cracked windscreen, or because the engine ran rough after the owner decided to fuel the engine with rubbing alcohol. None of that means the Ferrari is poorly designed, much less that it is not designed at all. In the same way, the best explanation for the origin of the human jaw remains intelligent design, with the competing explanation of mindless evolution huffing and puffing far in the rear, struggling to provide more than the vaguest of imaginative stories as to how unguided natural processes could have fashioned such an engineering masterpiece.

8. The Middle Ear

Two Views of the Middle Ear

"The ear is a remarkably sensitive and selective hearing organ," comment acoustics experts Stephen Elliott and Christopher Shera in the *Journal of Smart Materials and Structures*. "The human ear can detect sound that gives rise to internal motions of 0.3 nm, barely above Brownian motion, yet has a dynamic range of about 120 dB (decibels) and a resolution of about 0.5 dB."[1] The two authors aren't finished gushing. They continue:

> Young people are able to hear over a frequency range of about 10 octaves, with a frequency resolution of about 0.3 percent of an octave. This would seem to imply an extremely resonant system, and yet we are also able to resolve timing differences, of less than 1/100 of a cycle between sounds presented to the two ears, which helps us to localize sounds. All this is achieved with a sensing cell, the inner hair cell, that in isolation has a dynamic range of only about 30 dB and an untuned frequency response. It has been estimated that for all its impressive properties, the power consumption of the cochlea is only about 14 μW. Although our hearing is significantly enhanced by processing of the neural signals, it is the structure of the cochlea that provides a mechanical response that is the first step towards achieving this amazing performance.[2]

This is glowing praise for the design of the human ear. The organ is a tiny instrument and yet extremely sensitive and requires minimal power to operate. The Elliott/Shera paper explains that in some regards the ear operates at the limit of what is physically possible and far beyond what human engineers can produce.

It is sometimes wrongly assumed that human hearing is not highly optimized because animals such as dogs can hear sounds that we cannot (e.g., quiet sounds or high-pitched sounds). In fact, humans have all-round hearing capabilities that are just right for human activities. And studies have shown that humans actually have superior hearing to animals in terms of differentiating among very similar sounds. Such a capability is crucial for activities like human speech, where different syllables can have very similar sounds. For example, a 2018 paper in *PNAS* reports the following findings: "Combined with human behavioral data, this outcome indicates that the frequency analysis performed by the human cochlea is of significantly higher resolution than found in common laboratory animals." The authors then add this frank admission: "This finding raises important questions about the evolutionary origins of human cochlear tuning, its role in the emergence of speech communication, and the mechanisms underlying our ability to separate and process natural sounds in complex acoustic environments."[3] In other words, scientists do not understand how the human ear can be so sensitive at differentiating similar sounds, much less how it evolved.

Undeterred, evolutionists assert that the ear evolved from a relatively primitive synapsid reptile ancestor.[4] Riley Black, in describing this scenario in *Smithsonian Magazine*, says the synapsid reptile "jaw bones closest to the back of the jaw became smaller and specialized to transmit vibrations to the ear, improving synapsid hearing. The malleus, incus, and stapes of our inner ear are the remnants of these ancient jaw bones."[5]

But as we'll see, this explanation is wholly inadequate.

A Masterpiece of Acoustic Design

Figure 8.1 shows the layout of the mammalian ear. The outer ear collects sound waves and funnels them to the eardrum, which causes it to vibrate. The middle ear has a chain of three tiny bones, the malleus (hammer), incus (anvil), and stapes (stirrup), collectively called ossicles. These transmit vibrations from the eardrum to the cochlea in the inner ear, which contains the sensory organ for hearing.

When the vibrations reach the cochlea, they cause pressure waves in the fluid inside a spiral-shaped structure. In the cochlea, specialized receptor hair cells detect the pressure waves by converting the liquid motion into electrical signals that travel to the brain to be interpreted as sounds. Fluid vibration is ideal because it has enough inertia to move hairs efficiently. The spiral shape is ideal because it allows mapping of different frequencies of sound onto positions along the cochlea.

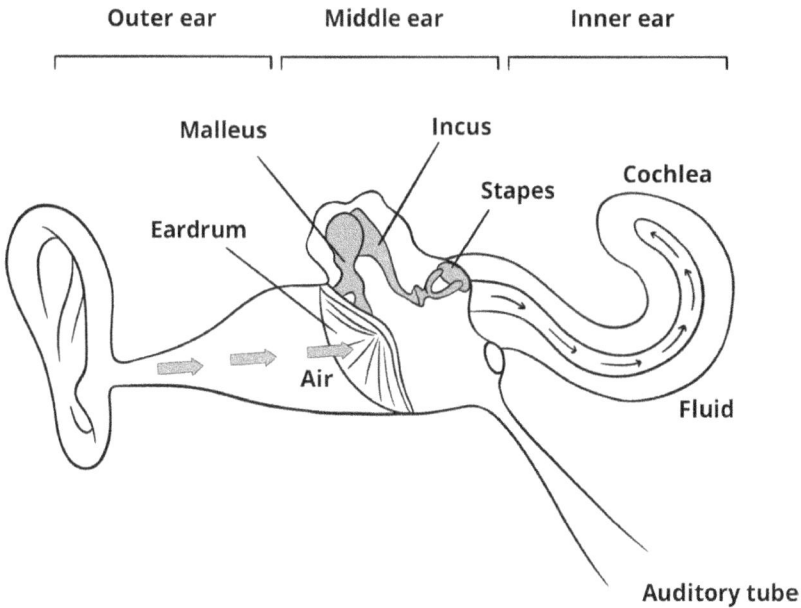

Figure 8.1. The outer, middle, and inner ear.

Precisely transferring vibrations from the eardrum to the cochlea is tricky. The ossicles form a kinetic chain that transmits vibrations. They must precisely amplify pressure through a lever system and protect the inner ear from loud vibrations. There is also an auditory (eustachian) tube that helps equalize the air pressure on both sides of the eardrum by allowing outside air to enter the middle ear from the nasopharynx (upper part of the pharynx).

Figure 8.2 highlights the middle ear's key ligaments and muscles. Some think of the middle ear as containing just three bones, but that is an oversimplification. The ligaments provide crucial support and stabilization while allowing the ossicles to move and rotate. And the muscles have the important task of fine-tuning the stiffness of the bone assembly.

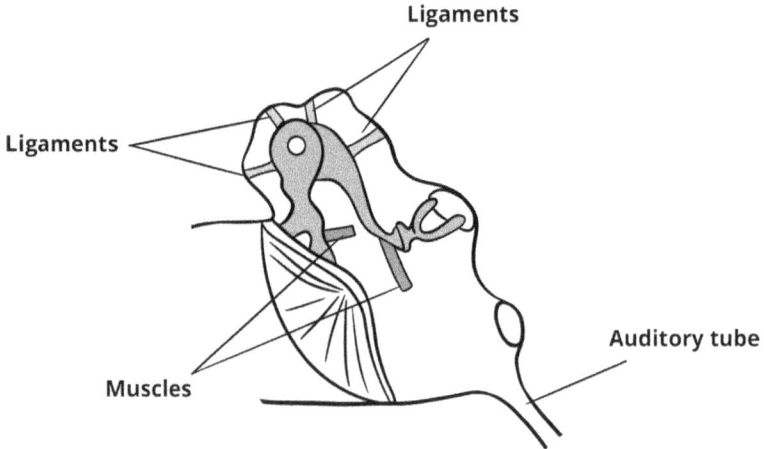

Figure 8.2. Ligaments and muscles of the middle ear.

The middle ear is an exquisite assembly of tiny precision components. The bones and muscles of the middle ear are the smallest in the body. The stapes bone is the smallest bone of all, just two to three millimeters long. A unique feature of the ear bones is that they do not grow during childhood. This is not surprising, because the design is so delicate that growing could misalign the precise arrangement.

Figure 8.3 shows how the three bones form a linkage system to convert pressure waves at the eardrum into pressure waves within the cochlea. Biology books almost always refer to the bones as a simple lever system, but this is another oversimplification. A simple lever system would have just one hinge, but the middle ear has four hinges, if you include the connection of the eardrum to the malleus and the stapes to the oval window of the cochlea. The exact dynamics of the linkage system are highly intricate and remain under investigation.[6]

The hinges of the bones are formed by the bone joints and tiny ligaments. The linkage mechanism has various functions. It serves as a lever for amplifying force; it lends elasticity to aid resonance; and it allows for the acoustic reflex, discussed below. The linkage system increases force by a factor of around 1.3. Because the vibratory area of the eardrum is around fourteen times greater than that of the oval window, the sound pressure is amplified by a factor of around eighteen.

As a mechanical engineer, I am very impressed by the fact that the middle ear is suspended by ligaments. This is a smart engineering choice, since it makes for an extremely efficient linkage, with low friction and minimal energy loss.

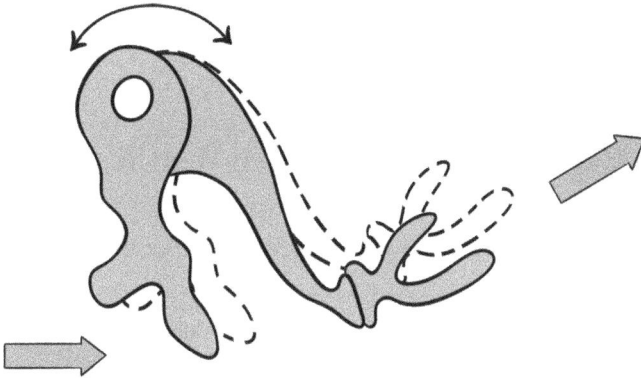

Figure 8.3. The bones that form a linkage system in the middle ear, converting pressure waves at the eardrum into pressure waves in the cochlea.

The Ingenious Acoustic Reflex

One consequence of having extremely sensitive ears is that they can be vulnerable to loud noises. When engineering sensitive sound-detection equipment, designers implement clever solutions to protect the device from loud noises. The human ear features a particularly clever solution.

The ear has an ingenious acoustic reflex that rapidly tightens the tiny muscles in the middle ear in response to loud noises. The stapedius muscle pulls the stapes of the middle ear away from the cochlea's oval window, while the tensor tympani muscle stiffens the assembly of ear bones by pulling on the eardrum and malleus. Together these actions dampen the transmission of vibrational energy to the cochlea when confronted with loud noises. This intricate safety device involves an astonishing level of precision design.

For those with normal hearing, the acoustic reflex threshold is about 70–100 dB. Typical sounds in this range are from machines like hair dryers and chainsaws. This helps explain why it is difficult to hear when you are near noisy machines. It's not just the noise of the machines; it's also the acoustic reflex dampening your hearing to protect your ears. Of course, the acoustic reflex cannot provide total protection from every loud noise. Very sudden noises such as gunshots are too fast for the reflex. Also, the reflex cannot protect against extremely loud noises or loud noises experienced over a long period. However, the acoustic reflex is a wonderful mechanism for protection against most of the loud noises people typically encounter.

An Ear Evolution Story: From Reptile to Mammal

Recall that evolutionists propose that the middle ear of mammals evolved from the middle ear of reptile-like creatures called synapsids. What would such an evolutionary transition have involved? Reptiles have a simple inner ear with one bone (the columella). The mammalian inner ear, as we saw, has a chain of three bones (malleus, incus, and stapes). Figure 8.4 shows how two of the jaw bones of reptiles supposedly migrated and rearranged themselves to become the inner ear of mammals.[7] The incus supposedly came from the reptilian upper jaw, while the malleus supposedly came from the reptilian lower jaw. There is no evidence such movements occurred; it is just speculation.

The scenario may seem plausible at first sight, but an engineering analysis reveals huge problems. The three bones of the mammalian ear are not just static bones but are joined in a chain with synovial joints. And each must be shaped very precisely to provide any appreciable benefit to the creature.

A second problem with the reptile-to-mammal picture is that it is a gross simplification of the middle ear in that it ignores the ligaments and muscles. There are at least nine components to the middle ear. So when evolutionists say they can explain the origin of two extra bones, they are ignoring at least six other parts, each precisely shaped and integrated into the whole.

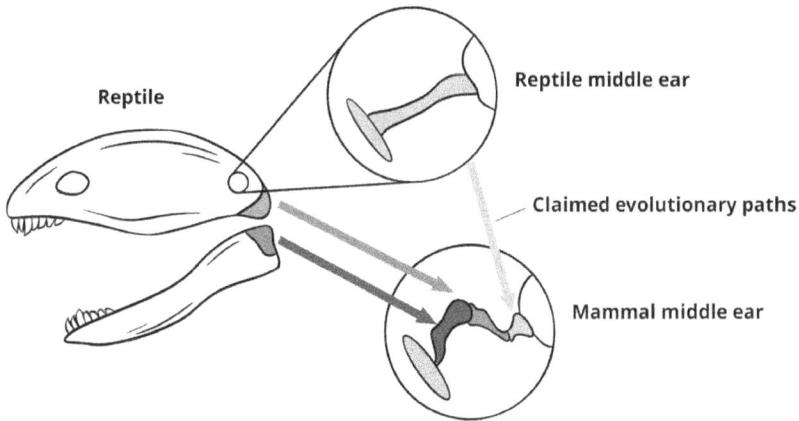

Figure 8.4. Supposed evolution of the reptilian middle ear into the mammalian middle ear.

Just the three bones themselves are irreducibly complex. As with the linkage in a car suspension, the linkage in the ear only works when all these parts are present, precisely shaped, and precisely assembled.

Such a system also requires many complex assembly instructions. When evolutionists say that "two bones migrated," they are not only ignoring other parts but are also skipping the explanation of how the precise assembly instructions arose to guide assembly of all these precision parts.

In sum, to say that the three bones "migrated and came together" ignores the fact that the three bones must be precisely shaped and precisely and intimately integrated with various specialized muscles and tendons via highly specified assembly instructions.

Yet another challenge for evolution should be mentioned here. We also possess a remarkable reflex that reduces the perception of sound when we are vocalizing. This helps us to be less distracted by external sounds when speaking. The stapedius reflex, triggered in anticipation of one's speaking, reduces the intensity of sound that reaches the inner ear by some twenty decibels. This design detail is hard to explain by evolution. How can it be a survival advantage to have the convenience of being able to speak with slightly less distraction? However, such a fine detail is readily accounted for from within an intelligent design paradigm.

How Pictures Can Hide Intricate Engineering

Verbal and visual descriptions of proposed evolutionary stages can be deceptive. Discussions of the middle ear are a prime example. Evolutionists use words like "migrated," "shrank," and "changed shape" when describing the supposed evolution of the three ossicles. And the accompanying illustrations typically leave out knotty engineering details indispensable to the middle ear.

There is a world of difference between static pictures and dynamic three-dimensional parts that interact in a complex assembly. The middle ear is not a two-dimensionally static picture but an immensely complex three-dimensional moving system with many critical design details. Pictures cannot effectively illustrate the precision interfaces between the bones or the complex assembly of linkages and joints.

Consider an analogy. Someone could present a diagram (Figure 8.5) arguing that a sundial could evolve into a Rolex watch through a series of random natural events. After all, how different are they really? The gnomon (pointer) could split and become the hour, minute, and second hands. The dial plate could become the watch face. But the diagram obscures an enormous amount of engineering inside the watch that allows it to function as it does.

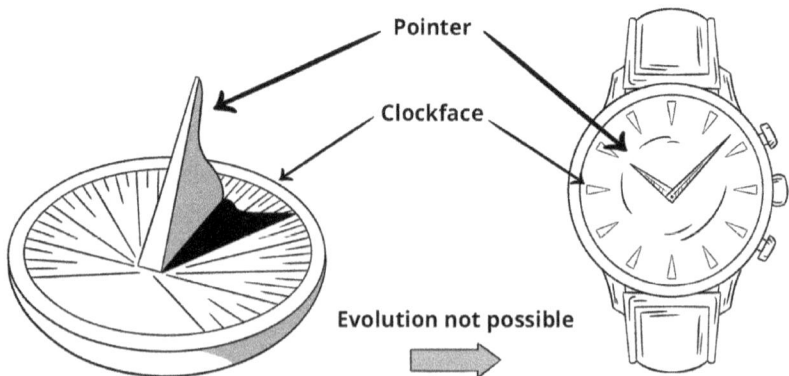

Figure 8.5. Sundial-to-watch evolution.

Consider the precision gears inside a Rolex watch (Figure 8.6). These obviously could not evolve from a sundial. Those gears were hidden in the pictorial representation of the evolution of a Rolex watch in Figure 8.5. In the same way, books that use pictures to illustrate evolution commonly leave out much or all of the precision-engineering details.

Figure 8.6. Precision gears inside a watch.

Lest anyone imagine that a Rolex watch is surely more complex than the middle ear, I should underscore that just the opposite is the case. Humans can make Rolex watches; we are far from being able to construct even a crude but passable replica of the middle ear, much less one that, as part of a larger system, can build itself during embryological development.

Even my overview above—which pushes past the cartoonish oversimplifications that evolutionists proffer when advancing their reptile-to-mammal middle-ear scenario—remains a gross over-simplification. To even begin to avoid this would require an entire book dedicated to the ear, and one accompanied by meticulously animated video simulations.

The Evidence Is Loud and Clear

The performance and precision engineering of the ear has astounded engineers and scientists. Experts in acoustics such as Stephen Elliott and Christopher Shera have described how the ear functions at the limit of what is physically possible. Such precision engineering is an extremely awkward fit for an evolutionary paradigm that sees the mammalian middle ear as something jury-rigged by mindless processes from a simpler reptilian ear. But such precision is not at all surprising under the intelligent design paradigm.

9. THE EYE

Two Views of the Human Eye

During my time designing spacecraft systems with the European Space Agency (ESA), I had the honor of working on the solar array deployment system for the Hubble Space Telescope. It was a great experience, with NASA and ESA working together (although there was a running debate about which units, English or metric, we should use). During this project I saw firsthand how brilliant engineering can produce a powerful vision system. Yet the Hubble Space Telescope does not begin to compare with the vision system of the human eye.

Computer scientists M. Ponnavaillo and V. P. Kumar work on artificial vision systems. "The human visual system is a remarkable instrument," they write. "It features two mobile image-acquisition units. Each has formidable preprocessing circuitry, placed at a remote location from the central processing system (the brain). Its primary tasks include transmitting tiny images within a viewing angle of at least 140 degrees and resolution of 1 arcmin over a limited-capacity carrier, the million or so fibers in each optic nerve."[1]

That description only scratches the surface of how the eye is a masterpiece of engineering. In this chapter we will delve into just one aspect of eye design. Although by no means the most complex part, I have chosen it because it is in my area of expertise—the biomechanics of the eye. Most people do not associate the eye with biomechanics, but the organ is a biomechanical marvel, with the muscles of the eye among the hardest working in the body in terms of the number and speed of movements.

Despite the high praise that experts in the field have heaped on the design of the human eye—both its biomechanics and its many other aspects—evolutionists like Richard Dawkins, Nathan Lents, and Abby Hafer claim it is a poor design. As we shall see, their claims contradict well-known published scientific research.

Basic Operation of the Eye

Figure 9.1 shows the main parts of the human eye. Light first enters the eye through the cornea, the clear front layer of the eye. The cornea is shaped like a dome and helps direct light to the lens. Light then enters the eye through an opening called the pupil, located in the iris. The iris can control the diameter of the pupil to optimize the amount of light entering it for different conditions, such as brighter or dimmer environments.

Eye Anatomy

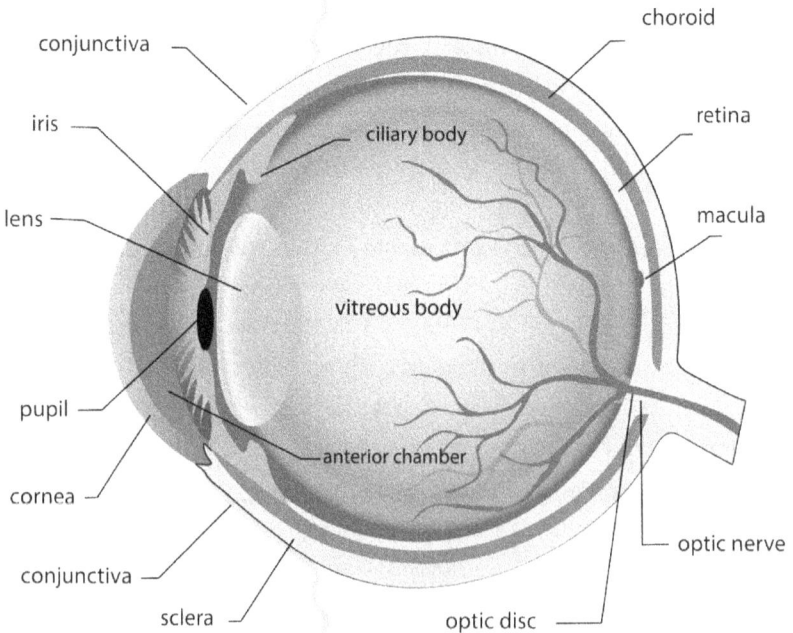

Figure 9.1. Eye anatomy.

Light then passes through an elliptical lens whose shape can be fine-tuned by several muscles to focus the light on the retina. The retina, a light-sensitive tissue layer at the back of the eye, contains thousands of specialized cells called photoreceptors. When light hits the retina, the photoreceptors convert the light into electrical signals and send them through the optic nerve to the brain, which processes the signals, turning them into images.

The sclera (the white of the eye) is the protective outer layer of the eye, containing mainly collagen fibers. The body of the eye is filled with vitreous humor, a clear gel-like substance that contributes to the intraocular pressure and helps maintain the shape of the eyeball. It also acts as a shock absorber, transmits light, and provides nutrients to the various structures within the eye.

A Most Capable Eye

Eyes have amazing capabilities. These include:

(1) Day and night vision

Human eyes contain millions of light-sensitive cells called rods and cones. Rods are optimized for low-light conditions such as nighttime. Cones are optimized for sharp color vision in daylight.

(2) Pupil reflex

The eyes optimize pupil size to adapt to different levels of brightness. Pupil size is rapidly adjusted by the iris muscles like a camera shutter system. Pupil size can vary from 2 to 8 mm.

(3) Distinguishing around 10 million different colors

For example, humans can distinguish between fine differences among the color of green plants.

(4) Eye tracking

The eyes are able to focus on objects even if the head is moving during activities like running, because of the rapid action of six external eye muscles that move the eyeball. The eyes move many thousands of times a day and yet they can function precisely the whole time.

(5) Rapid focusing

The eyes can focus rapidly due to the ocular muscles that change the shape of the lens automatically and allow proper focusing of light to the retina. They can change focus almost instantly and stay focused even when the head is moving around.

(6) Binocular vision

The brain combines the slightly different views from each eye, creating a single, three-dimensional image. This depth perception is important for tasks like judging distance, as when catching a ball or hunting wild game.

(7) The corneal reflex

The corneal reflex (eyelid blinking reflex) is an involuntary blinking of the eyelids triggered by stimulation of the cornea (such as by touching).

(8) Blinking

Blinking can be voluntary or involuntary. The eyes blink around 20,000 times a day. Blinking lubricates the eyes through corneal irrigation, using tears as a lubricant. Tears contain antibacterial substances that help fight infection and wash away dust and debris.

(9) Tear production

Other animals also produce tears for lubrication or when dealing with an irritant. But humans are the only animals known to cry tears. The physical act of weeping tears enhances the emotional feeling of sadness and communicates emotion while perhaps also providing physiological benefits in many cases. This has a parallel with goosebumps, which enhance the emotional feeling of sudden appreciation of beauty or the sublime.

(10) Self-healing

After injury the human eye can regenerate itself to some extent. For example, the outer layer of the eye, the cornea, can repair itself after getting scratched.

Biomechanics of the Eye

The six (extraocular) muscles that move the eyeball are shown in Figures 9.2a and 9.2b. The eye is roughly spherical and moves inside a spherical socket. The muscles can move the eye side to side (abduction-adduction) and up and down (elevation-depression), and they can rotate the eye around the axis of the pupil as shown in Figure 9.2b. Rotation of the eye is crucial for keeping the eye focused for mobile situations such as looking at an object when the viewer is walking or running.

The medial rectus is attached to the side of the eye next to the nose and moves the eye inwards while the lateral rectus is attached to the outer side of the eye and moves it towards the temple. The superior rectus muscle is attached to the top of the eye and moves it upwards. The inferior rectus muscle is attached to the bottom of the eye and moves it downwards.

The superior oblique is a particularly fascinating muscle. More on it below.

Incredibly, the eyes can move up to one hundred times per second, making them among the fastest muscles in your body. These rapid movements, called saccades, allow you to scan your environment efficiently and build a complete picture of your surroundings.

There are two other external eye muscles. These are responsible for eyelid movement. Internal muscles include the tiny sphincter iris muscles that open and close the pupils and the tiny ciliary muscles that change the shape of the lens.

Irreducible Complexity of the Trochlea Pulley

The trochlea pulley is my favorite part of the eye—a very clever design solution for improving the movement of the eye. Since muscles are anchored at the back of the eye, and since muscles only apply a force in contraction, the muscles can only pull the eye back. This constraint of approaching the eye from the back would restrict the range and efficiency of movement, but for a brilliant design solution to overcome this restriction: use of a ring-like device, the trochlea pulley, to allow the superior oblique muscle to approach the eye from the side, as shown in Figure 9.2b.

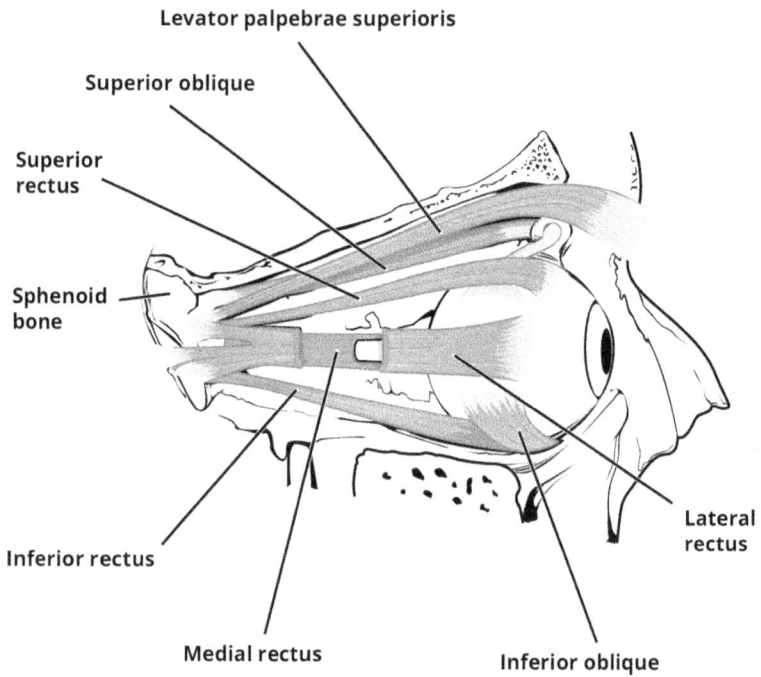

Figure 9.2a. Extraocular muscles of the eyeball.

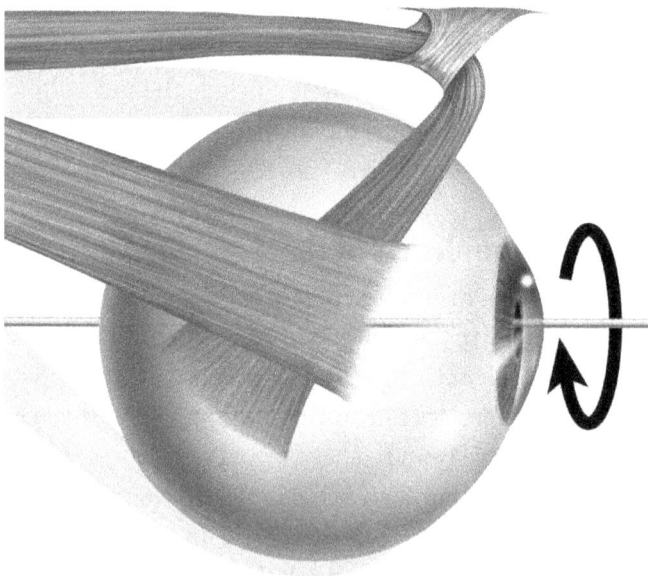

Figure 9.2b. Trochlea pulley.

The superior oblique muscle is located behind the eyeball, but the muscle passes through the trochlea pulley to allow the muscle to approach the eyeball from the side and hence pull the eye sideways.

This trochlea/muscle complex is not just a clever design feature; it is also irreducibly complex. The pulley and the muscle routing are both needed at the same time to be useful. For example, if the pulley suddenly appeared by accident, it would be no use without a muscle threaded through it. To propose that a pulley developed by accident and that a muscle somehow threaded through it at the same time is not credible.

The Iris: Better than Any Camera Shutter

When you look at eyes with lighter colors such as blue, you can see many tiny strings aligned in a radial direction, as shown in Figure 9.3, left. Most people do not realize that these string-like structures are precision muscles and marvels of engineering that can adjust the pupil diameter to within a few microns. In Figure 9.3, left, tiny lines can be seen around the inside rim of the iris. These are sphincter muscles, the actuators for reducing the size of the pupil. Slightly longer lines can be seen around the outside. These are dilator muscles, the actuators for increasing pupil size. For this to work, nerves must extend through the eye to control these iris muscles.

A modern camera shutter system for adjusting the amount of light into a camera is an impressive piece of engineering (Figure 9.3, right). However, engineers cannot make a shutter system as compact and accurate as the iris. And they cannot make actuators as small and accurate as sphincter muscles.

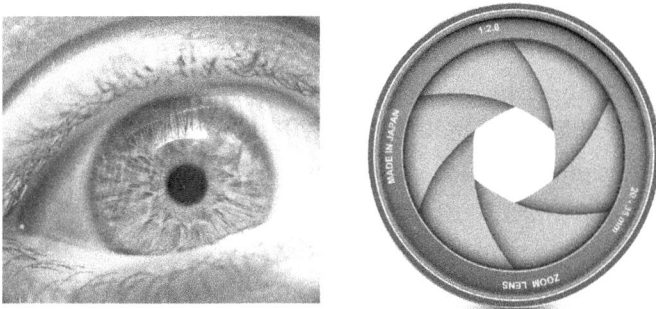

Figure 9.3. Optical shutter systems.

The Whites of the Eyes

One thing that enhances the human ability to make readily noticeable facial expressions is that the sclerae (the whites of the eyes) are clearly visible even from several yards away. This is not typically so with ape sclerae, which are generally darker, less exposed, and not particularly noticeable. In fact, humans are the only mammals where the sclerae are both typically white and prominently visible. (A functional benefit of mammals lacking prominent white sclerae is that this improves camouflage.)[2]

For humans, our prominent, white sclerae allow us to highlight certain emotions. For example, when our eyes open wide in surprise, the large white areas are exposed to a greater degree. When we squint while concentrating or expressing suspicion, the sclerae are much less visible. The sclerae also make it easier to notice when someone's eyes are rolled or shifted, actions that can communicate emotion or a shift in attention. In cases where the other person is a good way away, the whites of the eyes make it easier to tell if the person is looking at us, aiding communication.

The fact that humans have such visible sclerae is comfortably accommodated by design theory, since it makes good sense on ID grounds that humans were designed for high intelligence, moral sensibilities and, accompanying these, the capacity for uniquely nuanced social interactions enhanced by the ability to convey a variety of emotions through our eyes. Another reason the prominent whites make good design sense: Humans' higher intelligence makes them the apex predator in most circumstances and thus in less need of camouflage than are most other animals.

According to evolutionary theory, once upon a time one or more genetic errors gave an ape-like ancestor visible whites, a mutation that lent the ape-like ancestor and its offspring either survival or reproductive advantages, which were seized on by natural selection. Since human sclerae are both whiter and more exposed, it seems likely that more than a single point mutation is involved. In any event, the immediate impact of such a change would have been to make the ape-like creature look odd to the other members of his group and make it more difficult for the mutant to camouflage. Neither effect seems well calculated to enhance survival and reproduction.

Alternately, perhaps the feature evolved after our ape-like ancestors had grown considerably more intelligent and social, and the change was a hit with the opposite sex. It's an intriguing story, but it's important to keep in mind that such narratives are little more than airy conjecture stacked upon airy conjecture. It is easy to forget this because museums and textbooks sometimes present eyes with white sclerae in realistic-looking recreations of so-called "missing link" ancestors.

Earlier we saw how the ape-like creature Lucy, an *Australopithecus afarensis* hominid, is regularly given an upright stature in museum exhibits and artist renderings despite a lack of cogent scientific evidence for this. Lucy is also given white, prominent sclerae, another human-like feature. But there is even less evidence for this feature than for Lucy's claimed bipedalism. After all, the fossil record tells us nothing about the eyeball color of our ancient ancestors. What we can infer is that because Lucy probably was highly ape-like (see Chapter 1), she likely lacked prominent, white sclerae, since this is typical of apes. Depicting Lucy as having prominent, white sclerae might seem like a small detail, but it has a dramatic effect on the facial appearance and gives the impression that the creature was a transitional form, the bright light of an emerging human intelligence shining through.

Darwin's Doubt

This brief survey of the eye has focused on its biomechanics and barely hints at the layer upon layer of molecular intricacy required for the eye to function.[3] Darwin famously conceded that he understood why people would find it hard to believe that the eye could have evolved in the way he proposed. He addressed it under the subheading "Organs of Extreme Perfection," and as he put it, "To suppose that the eye with all its inimitable contrivances for adjusting the focus to different distances, for admitting different amounts of light, and for the correction of spherical and chromatic aberration, could have been formed by natural selection, seems, I confess, absurd in the highest degree."[4]

He went on to suggest a few possible steps in a hypothetical evolutionary pathway from light-sensitive patch to fully formed eye. But remember, Darwin and his contemporaries were unaware of the true complexity of the eye at both the macro and micro level. The idea that the eye could have arisen through a series of small, random variations, each step fully functional, and in this way become more sophisticated than any camera technology ever devised by human engineers, is an idea far more absurd today than it was in Darwin's time, given everything we now know about the eye's unparallelled sophistication. Biochemist Michael Behe emphasizes the point memorably:

> Now that the black box of vision has been opened, it is no longer enough for an evolutionary explanation of that power to consider only the *anatomical* structures of whole eyes, as Darwin did in the nineteenth century (and as popularizers of evolution continue to do today). Each of the anatomical steps and structures that Darwin thought were so simple actually involves staggeringly complicated biochemical processes that cannot be papered over with rhetoric. Darwin's metaphorical hops from butte to butte are now revealed in many cases to be huge leaps between carefully tailored machines— distances that would require a helicopter to cross in one trip.[5]

Answering Claims of Bad Design

It has long been a talking point of evolution defenders that the vertebrate eye is wired "backwards," that is, in a way that no savvy engineer would have chosen but that the higgledy-piggledy evolutionary process might well have.

In his 1986 bestseller *The Blind Watchmaker*, Richard Dawkins opined:

> Each photocell is, in effect, wired in backwards, with its wire sticking out on the side *nearest* the light.... This means that the light, instead of being granted an unrestricted passage to the photocells, has to pass through a forest of connecting wires, presumably suffering at least some attenuation and distortion (actually probably not much but, still, it is the *principle* of the thing that would offend any tidy-minded engineer!).[6]

Twenty-three years later, after his error had been pointed out numerous times, Dawkins was still peddling the same misinformation. It "is not just bad design," he thundered, "it's the design of a complete idiot."[7]

Biologists George Williams, Kenneth Miller, and Douglas Futuyma have lodged similar complaints against the eye.[8] Hafer joined the chorus more recently. "The first bit of bad design in our retina is that there are blood vessels sitting on the surface of the retina, blocking the light," she writes. "A decent designer could have put those blood vessels behind the retina. The second bit of bad design is that there are nerve fibers that also sit on the surface of the retina, blocking even more light. These, too, could have been placed behind the retina, if anyone had been trying to design the eye properly. Unfortunately, no one was."[9]

Jonathan Wells notes that already by 1986 and Dawkins's "tidy-minded engineer" critique of the eye, they should have known better. Wells's summary of the matter is worth quoting at length:

> If the rods and cones were to face the incoming light, as evolutionists claim they should, the blood-filled choriocapillaris and the RPE would have to be in front of the retina, where they would block almost all of the light. By contrast, nerve cells are comparatively transparent, and they block very little of the incoming light. Because of the high metabolic requirements of rods and cones and their need to regenerate themselves, the inverted retina is actually much more efficient than the "tidy-minded" design imagined by evolutionary biologists.
>
> The blind spot is not a serious problem, first of all because it is so small and second of all because the blind spot produced by the left eye is not in the same place as the blind spot produced by the right eye. This means that in humans with two good eyes, the field of vision of one eye covers for the blind spot of the other eye, and vice versa.
>
> Most of the research cited above documenting the essential functions of the choriocapillaris and RPE was published before 1986. But Dawkins, Williams, Miller, Futuyma, and Coyne didn't bother to check the scientific literature. They simply assumed that

evolution is true and that they knew how an eye should be designed. Then they concluded that the human eye is badly designed, claimed it as evidence for evolution, and ignored the contrary evidence. This is zombie science at work.[10]

It gets worse. In 2015 Erez Ribak reported in *Scientific American* on a new finding. "Until recently it seemed as if the cells in the retina were wired the wrong way round, with light travelling through a mass of neurons before it reaches the light-detecting rod and cone cells," he wrote, setting the scene in what can best be described as an overly generous reconstruction of the evolutionists' prior insistence on the backward wiring. Ribak continues: "New research presented at a meeting of the American Physical Society has uncovered a remarkable vision-enhancing function for this puzzling structure."[11] Specifically, researchers from the Israel Institute of Technology found that dense glial cells (also called Müller cells), which span the retinal depth and connect to the cones, guide light—just like fiber-optic cables—and hence improve vision.[12]

Thus, what some had claimed was a design weakness (backward wiring) that pointed to unguided and suboptimal evolutionary design is now known to be a design strength on multiple levels. The researchers started publishing their findings in 2010.[13] By 2015 their research was well established and attracting widespread publicity. Despite this, evolutionists like Lents and Dawkins continue to claim that the human eye is poorly designed due to being wired backwards. "Light must travel through a thin film of tissue and blood vessels before reaching the photoreceptors, adding another layer of needless complexity to this already complicated system," asserts Lents. "To date there is no workable hypothesis that explains why the vertebrate retina is wired backwards."[14]

Lents published this statement in 2018, three years after the prominent *Scientific American* article explaining another sound functional reason for the "backward" wiring, and a year after Jonathan Wells highlighted the findings in his 2017 book *Zombie Science*. Once again Lents, like Dawkins, Miller, and Futuyma before him, is guilty of following his evolutionary paradigm rather than the scientific evidence.

In her book *The Not-So-Intelligent Designer*, Hafer was very derogatory towards intelligent design scientists, claiming that "ID hates biology."[15] And yet it was Hafer who consistently ignored scientific facts throughout her book in order to follow an evolutionary paradigm. Hafer published her book in 2016, a year after the "wired backwards" claim of bad design was debunked in *Scientific American*.

Anyone Can See the Evidence

The plain evidence, increasingly insistent with every new discovery revealing new layers of sophistication, is that the eye is an absolute marvel of engineering and beyond anything that human engineers can produce. Given that engineers are far less constrained than the evolutionary process, the superiority of the eye to human technology strongly suggests that it did not evolve via mindless evolutionary processes, but instead was what it appears to be—the work of a designing genius.

10. SKIN

Two Views of Human Skin

"Human skin is remarkable and, in several ways, unique," writes anthropologist Nina Jablonski. "It is mostly naked, it is sweaty, it is tough yet sensitive, and it comes in a range of colors. The plentiful sweat glands in human skin secrete watery fluid that evaporates and helps to cool us when we are hot, and the outer surface of our skin is rich in specialized keratin proteins that help the skin to resist physical and chemical insults while remaining exquisitely sensitive to touch."[1]

Skin, in other words, is remarkable for combining toughness with sensitivity, and for packing so many functions into such a thin structure. Its density of functions has astonished scientists and engineers alike. Oxford University research fellow Monty Lyman dedicated an entire book to the subject: *The Remarkable Life of the Skin*.[2] There he describes some of the astonishing capabilities of skin, such as its communicating with our brains to enable us to interact safely, enjoyably, and precisely with our surroundings.

I have served as dissertation advisor for PhD students who spent four years developing some aspect of an artificial skin material for robots, and from this I witnessed the immense challenges involved in designing artificial skin. Engineers have spent millions of dollars trying to develop artificial skin for robots but have not come close to matching the performance of human skin. They cannot make sensors as small as those in human skin, especially the pressure sensors. And they cannot reproduce the elastic properties of skin. Engineers are simply in awe and humbled when they study human skin.

Despite the amazing design of skin, evolutionists claim that humans are just "naked apes" with evolutionary leftovers such as goosebumps. This chapter will highlight some of the impressive design features of human skin and respond to evolutionary claims.

The Anatomy of Human Skin: A Miracle of Design

Our skin is a fantastically sophisticated composite structure. There are over 150 complex proteins in skin. The main types are collagens, extracellular matrix proteins (elastin), keratins, and cellular proteins such as myosin. Some are structural proteins, such as keratin. Others have specialized functions, like helping produce vitamin D.

The skin has three layers—epidermis, dermis, and hypodermis. The epidermis is the outer layer and is less than 1 mm thick. The dermis has an average thickness of around 2 mm. The hypodermis (also called the subcutaneous tissue) is the deepest skin layer, situated closest to the muscle.

The epidermis layer has keratin, whose keratinocyte cells interlock like an artfully constructed stone wall to form a tough outer layer. Keratin is a very fibrous protein, ideal for creating a hard-wearing surface. The keratinocytes and the surrounding fatty mortar also make our skin waterproof.

The dermis contains collagen and elastin, which lend skin its shape and elasticity. The subcutaneous layer is mostly fat but far from useless. It provides structural support to the skin and functions as an insulator, particularly useful in cold weather. It also absorbs shock when the body is struck, minimizing damage to internal organs.

Skin thickness varies according to the needs of the body. Around the eyelids the skin is only 0.5 mm thick, whereas on the palms of the hand and soles of the feet, the thickness is around 4 mm. The heels of the feet also have a plantar fat pad of soft tissue between the skin and heel bone. The pad's honeycombed structure of fibrous, elastic chambers is filled with fat globules and serves nicely to absorb shocks, spreading the pressure across the surface of the heel.

The dermis of the body contains millions of blood capillaries, millions of sweat glands, and millions of sensors, including pain sensors, temperature sensors, and touch sensors. The dermis also has countless immune cells to prevent infection.

The skin feature I find most amazing is the design of individual body hairs. The human body has several million body hairs. Each hair has its own tiny muscle, lubricant gland, blood supply, and nerve endings close to the base of the follicle. These nerve endings are activated when the hair is deflected. The nerve endings can detect the tiniest of deflections caused by small objects such as tiny insects.

Researchers have discovered that there are different hair types that have different sets of nerve endings, which means different hair types detect different types of touch sensations. Also, each hair type is evenly spaced and patterned throughout the skin for an effective sense of touch.[3]

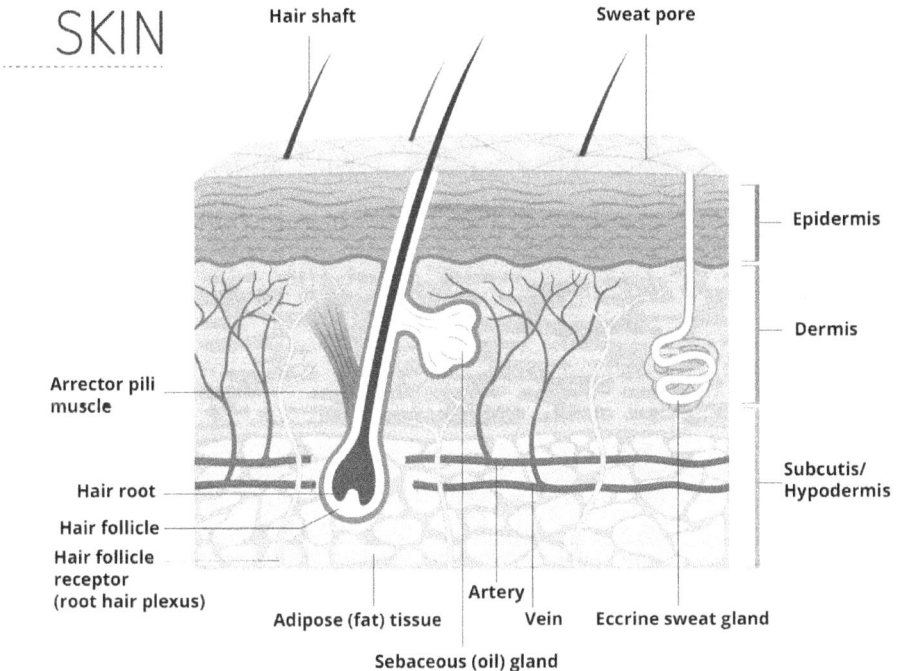

SKIN

Hair shaft

Sweat pore

Epidermis

Dermis

Arrector pili muscle

Subcutis/ Hypodermis

Hair root

Hair follicle

Hair follicle receptor (root hair plexus)

Adipose (fat) tissue

Artery

Vein

Eccrine sweat gland

Sebaceous (oil) gland

Figure 10.1. Anatomy of human skin.

Smart Self-Renewing Skin

Another impressive feature of human skin is that it can maintain its sensitivity even after decades of hard use. The skin of the whole body is completely renewed about once a month. In an eighty-year lifetime the skin is replaced around a thousand times. In old age the skin can still have a fine sense of touch. Robot engineers can only dream of creating a material that can reproduce itself. That is why robot parts wear out so quickly compared to the human body.

Keratin cells are generated from stem cells deep in the epidermis, where they gradually move up through the epidermis. Incredibly, skin not only has the ability to continually replace itself, but creates new skin at an optimal speed. The skin responds to excessive wear by replacing itself at a faster rate.

Research has even found evidence that skin cells have clocks that run on a 24-hour cycle.[4] Overnight, skin cells are created rapidly, preparing the epidermis for the sunlight and wear of the next day. During the day, skin cells then selectively switch on genes involved with protection against the sun's ultraviolet rays.[5]

It can be tempting to think of our skin as a simple covering. In reality it is a smart living system that intelligently maintains and repairs itself.

The Swiss Army Knife of Organs

Researchers often praise human skin for its multifunctionality.[6] When I am grading student design projects or assessing the design of an industrial product, I always look to see how multifunctional it is. A common hallmark of top designers is their ability to cleverly wring multiple functions from a single material or system. When I look at the multifunctionality of human skin, I see the fingerprints of a designer of unmatched intelligence. The introduction of this chapter touched on the skin's many functions. What follows is a more detailed overview, grouped into general areas.

(1) Covering functions

The skin envelopes the body, protecting it from impacts and scratches, harmful chemicals, ultraviolet radiation, and microbial threats such as bacteria. The skin is a dynamic barrier because of how it is stretched when joints move. At joints like the elbow, knee, and fingers, the skin is stretched millions of times in a lifetime of eighty years, and yet it still functions as a covering. No human-made material can begin to match that performance.

(2) Thermal functions

Human skin contains sweat glands, which help with thermo-regulation and excretion of waste. Other animals have sweat glands, but humans have a uniquely large number of them (650 per square inch), capable of pumping out several liters of sweat per day to cool the body. Because of the principle of heat of vaporization, the body is able to offload a lot of heat as sweat evaporates from the skin. (Another reason humans enjoy outstanding temperature control compared to other mammals is that we can adjust our clothing. If a polar bear sprints to catch a seal, it can't remove its coat, and it will take a long time to cool down. In contrast, humans can cool down faster by adjusting clothing.)

(3) Lubricating/antimicrobial functions

Human skin contains millions of tiny, specialized glands that pro-duce and secrete an oily substance (called sebum) that lubricates and protects the skin and hair. The ceruminous glands, located in the ear, produce earwax (cerumen). All these lubricating glands play a vital role in maintaining skin moisture and protecting against infections.

(4) Sensory functions

The skin contains a range of specialized touch receptors, called mecha-noreceptors, that detect many types of touch sensations. The Pacinian corpuscles, deep in the dermis of the skin, register vibration. Ruffini endings, also in the dermis, detect stretching of the skin. The Meiss-ner corpuscles, in the epidermis layer, are stimulated by skin motion. The Merkel cells, where dermis and epidermis meet, are optimized for detecting points and edges. These touch sensors complement the sense of touch achieved through body hairs. The skin also contains pain sensors and temperature (hot and cold) sensors.

(5) Vitamin D production

The epidermis is the major source of vitamin D for the body. Under the influence of ultraviolet radiation from sunlight, an organic compound in the epidermis called 7-dehydrocholesterol is converted to vitamin D.

(6) Facial expressions

The skin of the face can be stretched in various ways to create a myriad of facial expressions.

(7) Hair functions

One of the many differences between humans and apes is that humans have fast-growing long hair on the head and short hairs on the body. Long scalp hair has an insulating and aesthetic function. Body hair fulfills a touch-sensing function. It also produces the goosebump effect. (More on this below.)

Answering Claims of Bad and Vestigial Design

Despite the abundant scientific evidence that skin is an engineering marvel, evolutionists claim that human skin has design flaws due to its supposed evolutionary ancestry. According to anatomist Alice Roberts, a design flaw in human skin is that some people's skin is vulnerable to cancer. "Pale skin is prone to sunburn—which increases the risk of developing skin cancer,"[7] she writes. The first point to make in response is that human skin does have a degree of resistance to UV rays. It contains a photoprotective pigment called melanin that absorbs UV radiation and shields the epidermal-cell DNA from it. The amount of melanin in the skin depends on skin color, with darker skin having more protection. For people who do not have dark skin, it is necessary either to avoid exposing oneself to long periods of strong sunlight or to use sunscreen.

Evolutionists will argue that the design is substandard because it would be better if people with light skin did not need to take precautions with strong sunlight. But there are many areas in life where protection from environmental conditions is needed, as in very cold weather or dust storms. So taking protection from strong sunlight is just another one of those environmental precautions.

Some evolutionists also might argue that it would be better if all humans were dark skinned, since dark-skinned people don't sunburn as easily. This is certainly a great benefit of having darker skin. But the variety of human skin color types lends beautiful variety to the world. Having the whole human family be either all light skinned or all dark skinned would be far less aesthetically rich than the rainbow of human skin colors that actually exist. Additionally, the human body appears designed with a degree of adaptive capacity to the skin, partly in the individual but more dramatically in people groups over generations, such that those living for many generations in higher latitudes, where there is less intense sunlight, adapt lighter skin. It's a trade-off. Lighter skin isn't as good at protecting against prolonged, intense sunlight; but it's better at producing vitamin D, precisely what is needed in places far from the equator, since the kind of UV radiation that produces vitamin D in the skin reaches the surface of the earth less often the farther from the equator one goes. This adaptive capacity is itself a remarkable design feature.

But Roberts's view is that it's not good enough, and she suggests cephalopod (e.g., squid and octopus) skin, with its ability to quickly change color, as a superior alternative, one that presumably an intelligent designer would have chosen for the human form if running the show but that wasn't available to the blind evolutionary process when it was gradually converting a line of apes to humans. This suggestion for improving human design is no better than her others. Cephalopod skin, with its ability to rapidly change colors, would place massive energy demands on the human body and require an additional layer of neurological complexity on top of an already unusually complex neurological system. Cephalopod skin is also thinner and less durable than human skin. Nor is it clear how the cephalopod's color-changing apparatus could be integrated with human sensory nerves and sweat glands.

Design always involves compromises among the competing features the designer wishes to engineer. Besides the trade-off between robustness against intense sunlight (darker skin) and an optimal capacity for vitamin D production (lighter skin), there's this: Because human skin is extremely sensitive, it is inevitable that it will not be as tough as the skin of animals such as reptiles.

Another claimed design flaw in human skin is a reflex whereby body hair stands on end, yielding goosebumps. The scientific term for hair standing on end is piloerection. It is caused when thousands of tiny muscles (called the arrector pili) attached to hair follicles contract and raise the hairs. The way that each individual hair is thus actuated by a tiny muscle is a marvel of engineering. The reflex is caused by the sympathetic nervous system, which responds automatically to emotions. Our emotional brain responds more quickly to external stimuli than our cognitive brain, so we can get goosebumps before we even consciously register the cause for the goosebumps.

Piloerection can be caused by an emotional reaction to beauty, such as hearing beautiful music or seeing a beautiful sight. It can be caused by an emotional connection with a person, such as a romantic feeling. And it can be caused by an environmental experience, such as feeling a cool breeze.

Some evolutionists argue that it is a primitive evolutionary leftover. "It reminds us of a time when our ancestors had real fur instead of tiny hairs," wrote Abby Hafer, "and this reaction, though now functionless, is still with us."[8]

Functionless? Goosebumps have a readily discernable function: They enhance certain human emotions and, understood thus, make good design sense. Think about it. When people recall a highly emotional experience, they sometimes share that the experience gave them goosebumps. Undoubtedly, they share this detail because the goosebumps enhanced that experience. Reductionist utilitarianism may consider such a benefit beneath consideration, but then so much the worse for reductionist utilitarianism.

A Touch of Genius

Human skin is unique, very different from the skin of apes. We are not mere "naked apes." We have more sensitive touch sensors, far more sweat glands, and far smaller body hair. Additionally, our facial skin can be morphed into a myriad of expressions far more numerous and nuanced than is the case with any other animal. Human skin is an incredible multifunctioning organ. It not only performs many complex tasks but is also a beautiful covering. It is a brilliant example of ultimate engineering.

11. Birth Biomechanics

Two Views of Human Birth

In the *Annual Review of Biomedical Engineering*, researchers James Ashton-Miller and John Delancey, while acknowledging the reality of childbirths that go awry, marvel at one of the most extreme biomechanical processes in the human body. "Vaginal birth," they comment, "is a remarkable event about which little is known from a biomechanical perspective.... Computer models show that the stretch ratio in the pelvic floor muscles can reach an extraordinary 3.26 [326 percent] by the end of the second stage of labor."[1]

Such elasticity during the birthing process is crucial. Vaginal birth involves the baby being pushed out of the mother's body through the cervix and vaginal canal. This journey also involves getting through the opening in the pelvis and getting through a hole in the pelvic floor. A baby's head is typically about 4.4 inches in diameter. To get such a big object out of the womb requires big pushing forces and a lot of stretching in the birth canal.

In 2021 bioengineer Michele Grimm published a paper on the biomechanics of birth that describes the considerable forces involved in pushing the baby through the birth canal.[2] Grimm describes how it can require up to fifty pounds of force to push the baby during delivery. In extreme cases, the pushing force has been estimated at over one hundred pounds. These pushing forces come from a combination of contractions and the mother's pushing.

Despite the wonder of childbirth, evolutionists like Alice Roberts claim that human birth has design flaws due to its supposed evolutionary roots. This chapter will highlight some of the amazing biomechanics of vaginal birth and answer claims of bad design.

A Birth in Three Acts

I have witnessed five natural births, those of my five children. I must confess that in addition to experiencing the thrill of fatherhood and the concern for my wife during the arduous event, as an expert in biomechanics I also was fascinated by the engineering aspects of one of nature's most miraculous events. I was astonished to see how a woman's body is capable of going through such a rapid and dramatic change so as to eject a seven-pound baby. I was also surprised how quickly a woman can recover. My wife was fully mobile and active within 24 hours, which I found difficult to comprehend. As we'll see, all this is compliments of some astoundingly clever engineering.

For the bulk of this chapter we will home in on just two of the many clever engineering solutions required for human birth. But first, let's step back and take a quick look at the overall process. Birth can be broken down into three primary stages.

(1) Dilation of the cervix

The initial stage of birth is labor, during which the cervix gradually dilates. Dilation is broken into the latent phase and the active phase. In the latent phase the cervix dilates from 0 to 4 cm. In the active phase it goes from 4 to 10 cm. In a first pregnancy, the latent phase typically lasts around six hours. For subsequent births the latent phase can go much faster. During the active phase, the cervix typically dilates about one centimeter per hour in first pregnancies. For subsequent pregnancies, it's more like 2 centimeters per hour. However, it can be much faster, as was the case with my wife's deliveries!

(2) Expulsion of the baby

This stage begins at full dilation of the cervix and continues until birth, and it too is broken up into two subphases. The passive one involves the baby's head moving through the vagina. The active involves the mother pushing in step with the uterine contractions. The active part lasts around forty-five minutes for a first birth, though this can vary widely. After delivery, the umbilical cord is clamped.

(3) Placental delivery

This stage sees the delivery of the afterbirth (placenta and membranes). Stage 3 can last half an hour, though it can be hurried along with assistance. Breastfeeding may be encouraged during this stage since it seems to help the process along.

The coordination of all this is a masterful feat of engineering. In a precisely timed, multistage process, various hormones are delivered into the bloodstream by glands in the endocrine system to initiate major changes to the body. Of course, sometimes things go wrong, but it is remarkable how often things go right.

It is difficult to find an engineering analogy to vaginal birth. However, a key function of most space rockets is to deliver (give birth to!) satellites in space. As with vaginal birth, a rocket has to go through major reconfigurations in a short time span. The rocket must transition between different rocket stages during launch and then eject one or more satellites into orbit. Having worked on both of these extreme engineering challenges, I know from firsthand experience that they are very difficult to design. This has left me all the more in awe of the birthing process.

Pelvic Floor Muscles

The pelvic floor is crucial to the biomechanics of birth. This bowl-shaped structure, made mostly of muscle, forms a base for the pelvic cavity, which contains the bladder, the bowel, and (in the case of women) the uterus. There are two significant holes in the pelvic floor: the urogenital opening and the rectal opening. (See Figure 11.1.)

The urogenital opening is the gap at the front of the pelvic floor through which the urethra and (for females) the vagina and birth canal pass. The rectal opening, a centrally positioned gap, serves as the passage for the anal canal. During labor the pelvic floor muscles undergo a series of complicated changes to enlarge the urogenital opening to allow the baby through. (An additional accommodation for the birthing process: The pelvis in women is generally wider and shorter than in males.)

The pelvic floor includes both skeletal and smooth muscle, ligaments, and fascia. The ligaments and fascia are composed mostly of extracellular matrix, itself consisting primarily of proteoglycans and glycoproteins with collagen, along with elastin, which produces the extraordinary compliance and elasticity required for vaginal birth.

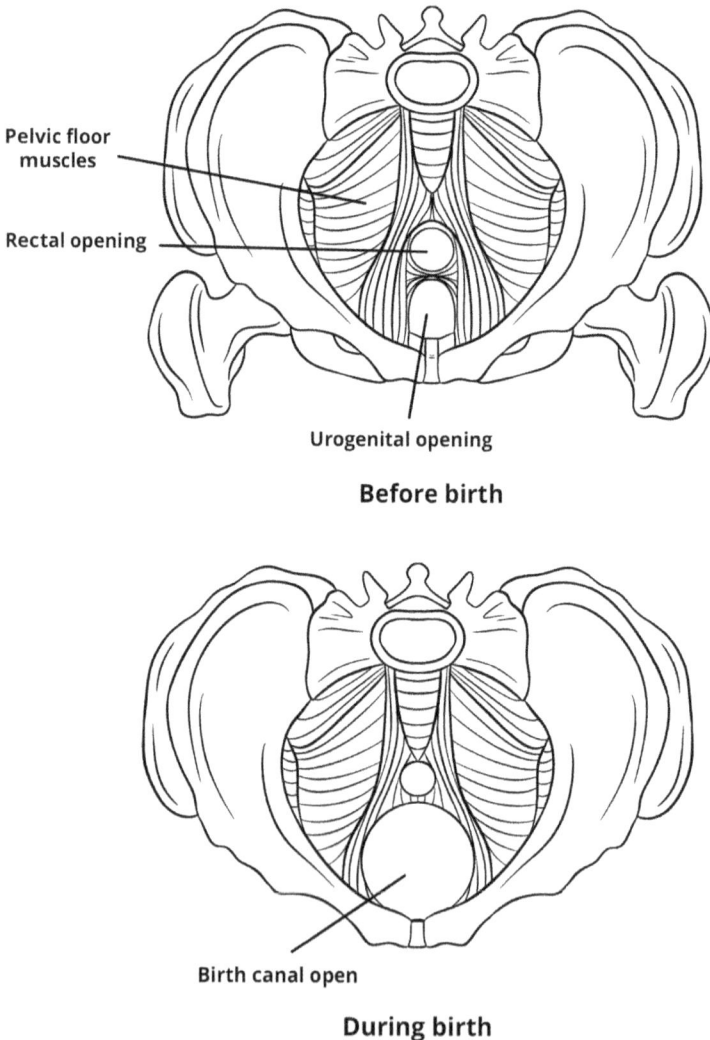

Before birth

During birth

Figure 11.1. Pelvic floor muscles, before and during birth.

Pelvic Floor Muscles:
Unsung Superheroes of Childbirth

During the birthing process, periodic contractions of the uterine muscles tighten the top part of the uterus, which pushes the baby downwards, putting pressure on the cervix. This downward pressure on the cervix slightly stretches the pelvic floor muscles.

During early labor, the blood vessels deliver the hormones progesterone and relaxin to the pelvic floor muscles, rendering the muscles and ligaments extremely pliable in order to allow the incredible stretching necessary for birth. The means by which these hormones render the muscles more pliable is not fully understood.

When the baby's head begins pressing on the pelvic floor, the mother's muscles relax further and the pelvic floor gradually opens the birth canal, guiding baby's head into proper alignment. When the cervix is fully dilated and baby's head is about to crown, the pelvic floor muscles must stretch near to their limit to make enough room for the head. The muscles also thin out at this time, and the considerable pressure exerted by the baby's head is enough to begin moving it through the birth canal.

At this point the mother is experiencing a strong urge to push, and she uses three sets of muscles to do so—the abdominal, pelvic floor, and hip muscles. These work with the uterus to push and guide the baby onward.

In the last stages of delivery, as the baby's head moves through the birth canal, the muscles of the pelvic floor reach their maximum extension of over 300 percent. Once the baby has been delivered, the sudden release of pressure frees the muscles to begin reverting to how they were before.

In my labs I have developed artificial muscle actuators using silicon materials and innovative geometries. At present, commonly available artificial muscle actuators cannot match the 300 percent extension of human muscles. Nor can artificial actuators match the compactness of human muscle.[3] Pelvic muscles are the unsung superheroes of vaginal childbirth.

Molding the Baby During Birth

Human birth requires not only enlarging the birth canal but also minimizing the circumference of the baby's head, a process called molding.[4] A baby has separate plates in the skull that allow for the preborn baby's skull to become squashed, and squashed without suffering damage, because the sutures of the skull have not yet fused, freeing the separate cranial bones to overlap. This reduces the diameter of the skull, allowing it to fit through the birth canal. A baby's skull is often misshaped immediately after delivery, but the skull regains its shape within about one week of birth.

Kick-Starting the Cardiovascular System

Just after birth, a baby loses the life-support system of the placenta and has to live autonomously. The baby will have to suck to get milk, swallow and digest the milk, and perform bowel movements. But before all this there is one very dramatic change needed instantly: kick-starting the cardiovascular system. I regard this process as one of the biggest miracles of engineering design in biology.

In the womb the baby's lungs contain fluid and are deflated. Oxygenated blood is delivered via the umbilical cord through a fetal circulatory system. This system contains dedicated features in the baby for placental life, such as a blood vessel (ductus arteriosus) and a hole in the heart (foramen ovale) that enables the blood to bypass the lungs. But then at birth there is the need to instantly change this fetal system to a completely autonomous blood circulatory system.

What triggers this monumental change is typically the baby's first breath. The first breath is usually a result of sudden exposure to air, which causes the lungs to inflate for the first time. This inflation removes the fluid from the lungs as well as drawing in air. The tiny air sacs (alveoli) in the lungs expand, and oxygen begins to diffuse into the bloodstream. This process is supported by the secretion of a special substance, surfactant, that reduces surface tension in the lungs, preventing alveolar collapse and enabling efficient breathing.

As the lungs fill with air, blood pressure increases, causing a flap to close over the ovale hole in the heart. In addition, the ductus arteriosus closes, redirecting blood flow to the lungs for oxygenation. Within minutes, the newborn's body transitions to a system where the lungs and heart work together to oxygenate and circulate blood.

For this transition to work, the ovale flap requires a precise design that makes it shut automatically due to the blood pressure change at birth. And closing the ductus blood vessel is also a complex action. If the ovale flap closes in the womb, the baby can die. And if the flap does not close at birth, serious problems occur. And yet it is remarkable how often the process works seamlessly.

Watching from the sidelines, it can seem that a baby adapts easily to autonomous life. However, inside the baby there are incredible feats of engineering making life outside the womb possible.

Answering Claims of Bad Design

The biomechanics of the birthing process are a marvel of making the seemingly impossible possible. At the same time, we should acknowledge that human birth is a fraught affair. In earlier times, without modern medical interventions, as many as 10 to 20 percent of births may have ended in the death of the child, either immediately or within the first week. Evolutionists insist that this checkered track record is due to some fairly elementary design flaws, errors that no savvy designer would have allowed but that are just what we could expect from that blind tinkerer, evolution by natural selection.

Evolutionists suggest that humans have it much worse than apes do when it comes to labor and delivery because women's physiology is a half-baked evolutionary revision of female ape physiology. Women have narrower hips than their quadruped counterparts, narrower hips being better for bipedal locomotion. And human infants have larger heads than ape infants, bigger to accommodate our bigger, more capable brains. According to evolutionists, if humans had been designed from the ground up, women would rarely have problems in childbirth. But because we are a catch-as-catch-can revision of ape physiology, the birthing mechanics of female *Homo sapiens* are sadly substandard.

This evolutionary explanation for why the birthing process has been a dangerous affair throughout most of human history has a superficial plausibility, but it runs into a couple of general problems. First, apes also experience failed deliveries, and while the rate may be slightly lower than for humans, it's not dramatically lower—with one study of zoo apes placing it at around 12 percent stillbirths.[5] Second, the ability for any complex structure to reproduce itself is a fantastically advanced process, one that the best human engineers are nowhere close to mastering. For asexual reproduction, the process is a wonder. For mammalian sexual reproduction, it is all the more astonishing. And anything this high-tech we can expect to be highly sensitive to perturbations.

A rough analogy is a baseball batter who regularly bats .800 against the best baseball pitcher of all time. Imagine a pitcher who combines the individual strengths of the best pitchers in baseball history—a hurler with a tireless arm capable of a 105 mph fastball, a wicked curve ball, and a fiendish change-up along with a passel of other sly pitches—sliders, cutters, an old-style knuckleball. You name it, each pitch is delivered with pinpoint accuracy. One could complain that the batter going up against this pitcher fails fully 20 percent of his at-bats. But anyone who knew something about baseball and the challenge of getting hits against even a moderately competent professional baseball pitcher would recognize that a hitter who could bat .800 against such a world-beating pitcher would be an unparalleled master of his craft, so good that he would probably have to be banned from major league baseball just to maintain rough parity among the teams.

In the same way, given the challenge posed by the big brains/big heads of human infants passing through relatively narrow hips optimized for bipedal locomotion (with its attendant advantages of making us more efficient long-distance travelers and freeing our hands to be optimized for the creation and use of tools), and given that this challenge is piled on top of the outlandishly high engineering demands of any mammalian birth, the fact that human mothers successfully birth their babies 80 percent or more of the time (absent modern medicine) is a testimony to some next-level design work that no human engineer is remotely close to achieving.

Any design for such a process will involve trade-offs, regardless of what story one prefers for explaining the origin of that process. As we saw in previous chapters, it is easy to pooh-pooh design choices where there are inevitable trade-offs, and cavalierly insist upon an alternate solution without taking a careful look at the trade-offs of that alternate strategy. But it's no way to do a competent engineering analysis.

Let's turn now to some specific complaints about the biological system of human labor and delivery, and see how these arguments for bad design hold up under scrutiny.

(1) The claim that a good design would not involve pain

Evolutionists argue that a proper design would have given women far bigger hips and pelvises, thus relieving them of much pain and difficulty during childbirth. Certainly, childbirth is challenging and painful. I have seen my wife experience the pain of childbirth five times and seen the intensity of the pain. And many mothers in labor have had it much worse. However, while giving women a significantly larger birth canal might reduce the pain of childbirth, it would make walking and running difficult. Because humans are bipeds and walk upright, the feet have to be close together so that they are under the center of gravity of the body. Having very wide hips would create angled legs that would not be good for bipedal locomotion.

The issue of pain in childbirth opens onto the larger question of why, if nature is the product of a good, wise, and powerful deity, is there any pain and death at all in the world? This is a profound question. It's also a theological rather than a scientific question, so it necessarily requires theological resources to answer. The Judeo-Christian tradition provides such resources,[6] which are often ignored or caricatured by professional atheists. Or worse, the atheist poses this theological question and then complains when a theological answer is given.

(2) The claim that damage to the pelvic floor in childbirth is proof of poor design

Sometimes the pelvic floor is damaged during childbirth. In some cases it takes a very long time to heal, or never fully heals. Evolutionists insist that this is clear evidence of a botched design. But lingering damage to the pelvic floor in childbirth, as it turns out, is in most cases in the same category as frequent lower back problems. That is, in many such cases, the pelvic floor is a victim of the modern lifestyle.

In technologically advanced societies, prolonged periods of sitting during work and leisure is the norm. In past ages, people spent most of their time on their feet with small periods of sitting. Now we have reversed that pattern—most people spend most of their time sitting and only short periods standing. It is not difficult to surmise that our maker optimized us for standing and moving about on our feet most of the day rather than for sitting around.

Too much sitting can cause tightness and weakness in the pelvic floor muscles. This is especially so if there is poor posture. Good sitting posture includes sitting with a straight back and shoulders and having knees at the same level as the hips. If someone does little exercise and sits too much, the pelvic floor muscles can cause health problems such as pelvic pain, loss of bladder control, backaches, and pelvic prolapse.

If a pregnant woman has weak pelvic floor muscles, she is more likely to suffer problems after giving birth, such as loss of bladder control. This is why women are often encouraged to do pelvic floor exercises. Women who do such exercises during pregnancy report less urine leakage in late pregnancy and in the first few months after birth. Pelvic floor exercise involves repeatedly contracting and relaxing the muscles that form part of the pelvic floor.

(3) The claim that it would be better to have a marsupial pouch

Evolutionists such as Abby Hafer and Alice Roberts have claimed that the female body would be a better design with earlier birth afforded by a marsupial pouch, since this would involve the delivery of a much smaller, easier-to-deliver baby.

"Now a wise Creator," wrote Hafer, "could have solved this engineering problem easily by doing something like—this!" Here she provides a picture of a kangaroo with her joey, and then elaborates: "The baby develops outside the body of its bipedal mother, in a nice comfy pouch complete with a nipple for nursing. Animals like kangaroos give birth to very small, embryo-like young that are placed in a pocket on the outside of the mother's body. This is where they continue their development. That's the way to do it if you're going to be a biped."[7]

This proposal falls under the easier-said-than-done umbrella. Human babies don't merely get bigger during their longer term in the womb. They take that time to develop vital organs to a healthy state. For example, babies born prematurely, before week 37, have a higher risk of developing lung problems such as infections. And the more prematurely they are born, the more likely they will suffer breathing problems. It's true that a marsupial pouch provides a more oxygen-rich environment for the developing joey than does outside air, but a human infant would still need to be dramatically re-engineered to be viable even in a pouch after just a few short weeks of embryological development. And this re-engineering would inevitably come with costly trade-offs.

While a marsupial pouch is a fine solution for the joey, the design is not appropriate for humans. The joey is born after thirty-four days and then spends over six months in the pouch. If one were to elect a placental mammal as a candidate for converting to a marsupial (mothers with pouches), humans would be far down the list of likely candidates, due to the requirements for properly developing our large, complex brains during early development. Put simply, the marsupial pouch is a poor candidate for the unparalleled demands of developing the most complex known organ on earth, the human brain.

Note, too, that even for the kangaroo and joey, there are costly trade-offs. The birthing process is easier for the kangaroo mother and joey, but the joey is more vulnerable to predation and death during its time in the pouch than a placental mammal is during its longer term in its mother's womb. In essence, the marsupial strategy trades one kind of mortality threat for another.

An evolutionist might even argue that the marsupial design is the "bad design," since whenever placental mammals have been introduced into biosystems characterized by native populations of marsupials, the newly introduced placental mammals typically outcompete the marsupials. But from within a design paradigm one isn't obligated to assume that a wise designer would wish to make every creature maximally competitive. The designer may have other creative virtues in mind, such as a greater variety of animal forms. After all, only a very dull soul would wish kangaroos and other marsupials into non-existence on the view that they often have trouble competing head to head with their placental peers.

Another reason a marsupial pouch is a less-than-ideal solution for humans is that it is inelegant for the human body. Yes, this is a more subjective argument. However, many of the most important things in life can be dismissed as "merely subjective" but should not be. Humans have a unique physical beauty in terms of proportions, curves, smooth skin, fine hair, and fine facial features. A pouch would compromise the beautiful form of the woman.

Nothing Ventured, Nothing Gained

Unquestionably, pregnancy and birth are risky. But hazards are a part of life, and a part of achieving anything great. Some of the most glorious achievements, such as the moon landings in the 1960s, involve great risks. Faulting the process of human birth because of the risks is like faulting an advanced space rocket design because launching humans into space is risky. Birth is an event with inherent risks, but as many a woman past and present would attest, the risk is worth it.

Human birth has many astounding engineering features, including extreme muscle stretching, extreme forces during the contractions and active pushing, unique hormonal communication, and skull molding and recovery. The process is yet another astounding example of ultimate engineering in human biology.

12. The Blood
Circulatory System

Two Views of the Circulatory System

"Nothing shows more clearly the perfect engineering of the heart than our own failed attempts to imitate it," writes Sian Harding, Emeritus Professor of Cardiac Pharmacology, Imperial College London. "This history of the total artificial heart is punctuated with both brilliant innovation and continual clinical failure."[1] Harding, past president of the European Section of the International Society for Heart Research, titled her article, "In Search of the Impossible Machine, the Artificial Heart." The reason for the term "impossible machine" is that Harding concluded from hard experience that designing an artificial heart as good as the human heart may well be impossible.

Despite high praise for the design of the heart by world experts in cardiology, evolutionists like Richard Dawkins and Alice Roberts insist that the heart has some glaring design flaws. As we shall see later in the chapter, their claims are not based on any scientific evidence or design principles.

The heart, of course, is only part of a larger and more complex system, one that itself is without parallel in the world of human technology. This chapter will explain why researchers describe this blood circulation system as "perfect engineering."

Our Circulatory System: An Ingenious Double Circuit

The human circulatory system circulates blood and lymph throughout our bodies. Mammals have an ingenious double system for circulating the blood, which features two major loops: the pulmonary loop and the systemic loop. The pulmonary loop circulates blood to and from the lungs to release carbon dioxide and collect oxygen. In the systemic loop, blood delivers oxygen and nutrients throughout the body and collects carbon dioxide waste.

This double-circuit design allows for two very different pressures. The loop to the lungs has a low mean arterial pressure of around 10–15 mm Hg, ideal for gas exchange through the thin vascular walls of the blood vessels. For the loop through the body the mean arterial pressure can be up to ten times higher, around 95 mm Hg, ideal for the long distances that have to be traveled.

To help circulation there are valves in the veins to prevent back-flow as blood is pumped back to the heart. Another important design feature of the system is the parallel piping of organs, which requires more complex plumbing compared to a single loop with everything in a series. But the extra blood vessels and joints of a double-loop system are well worth it, since they allow fresh blood to enter each major organ, and the system keeps pressure high for each of them.

The major parts of the circulatory system are the heart, blood vessels, and lungs, but the system is integrated with many other sub-systems, including other organs, the lymphatic system, muscles, and every other tissue in the body.

A Multifunctional Marvel

The circulatory system performs many complex functions, including the following:

(1) Collects oxygen

The lungs are where blood collects oxygen. There are millions of air sacs (alveoli) in the lungs, which fill with air when we inhale—air containing 21 percent oxygen. The oxygen passes through the very thin walls of the alveoli and into the millions of tiny blood-filled capillaries. Blood contains red blood cells, which are full of a unique protein called hemoglobin. The hemoglobin is specially engineered to readily bind with the oxygen. Incredibly, inside each red blood cell there are 200 to 300 million molecules of hemoglobin. The vast number of capillaries provides an enormous surface area, affording rapid and efficient gas exchange between the alveoli and the bloodstream. The oxygen-rich blood in the lungs is pumped back to the heart, ready to be pumped around the body.

(2) Collects nutrients and hormones

The intestines are where blood is topped up with nutrients such as sugars, fats, vitamins, minerals, and proteins. Digested nutrients are absorbed into the blood through millions of capillaries in the small intestine. The blood also collects hormones that are secreted from various glands in the body.

(3) Delivers vital chemicals to cells

Circulating blood transports oxygen, nutrients, and hormones around the body. When the red blood cells reach tissues that need oxygen, the oxygen is released from the hemoglobin and diffuses into the cells for energy. Nutrients such as amino acids, fats, glucose, and vitamins use carrier proteins to get into the cells. The circulation of hormones throughout the body allows organs to communicate. Hormones help control processes such as growth, metabolism, reproduction, and the operation of organs. Once a hormone reaches a target cell, it binds to cell receptors.

(4) Ferries material away from cells

Remarkably, the same blood that delivers nutrients also carries carbon dioxide and other waste away from cells to be removed from the body. As blood goes round the body from the heart, the oxygen content decreases and the carbon dioxide content increases. Carbon dioxide is expelled from the lungs. The kidneys filter urea and other waste from the blood, which then get added to the urine in the bladder. That's a simplified description. Other organs are also involved in processing toxins. The liver, for instance, processes a host of toxins and sends some (but not all) of those processed toxins back into the bloodstream to be filtered by the kidneys.

(5) Transports lymph fluid

The lymphatic system, a network of vessels and tissues, moves a watery fluid called lymph back into the bloodstream. Unlike the cardiovascular system, the lymphatic system is not closed. It maintains normal fluid levels in the body and absorbs fats so they can be returned to the bloodstream. The lymphatic system is a vital component of the immune system, working to protect against infection and destroy old and abnormal cells.

(6) Regulates temperature

Circulating blood plays a key role in temperature regulation by distributing heat throughout the body. The body can change the blood flow to the skin to control heat exchange at the skin's surface. If the body gets too hot, extra blood is sent to the skin to accelerate heat loss. If the body gets too cold, less blood is pumped to the skin to conserve heat.

(7) Fights infection and disease

Blood also protects against infection. In particular, white blood cells help fight infection and disease.

(8) Clots blood

Blood clotting reduces and stops bleeding when the skin is cut. Such injuries to the skin trigger a reaction in the blood vessels that activates platelets (specialized blood cells) and clotting factors (specialized proteins in the blood), which team up to retard and eventually stop the bleeding by means of a clot formed over the damaged area. This prevents the injured person from bleeding out while the body heals. When the injury is sufficiently healed, the clot is dissolved.

To perform so many functions in one circulatory system is astonishing from an engineering point of view. Engineers sometimes design fluid delivery systems that have to perform multiple functions. For example, diesel fuel systems often add both engine lubricant and cleaning fluid to the fuel, creating a fuel mixture that performs three functions. These fuel systems require many precision parts and fine control. But they are really simple compared to the body's circulatory system.

The Heart: An Ingenious Double Pump

There are more than 60,000 miles of blood vessels in the adult body. The heart has to generate enough pressure to pump through that entire length, so clearly it must be powerful and efficient. Additionally, in order to deliver blood to the two circuits in the body, the heart has to be a double pump with the two pumps precisely synchronized, no small order due to the big pressure difference between the two circuits.

Each pump needs an input and output chamber, so the heart has a total of four chambers as shown in Figure 12.1. Two chambers are called atriums, appropriate because that is where blood enters. The two exit chambers are called ventricles. These have thicker walls than the atria because the ventricles generate higher pressures to pump the blood. The atria are at the top of the heart and the ventricles at the bottom.

Anatomy of the Human Heart

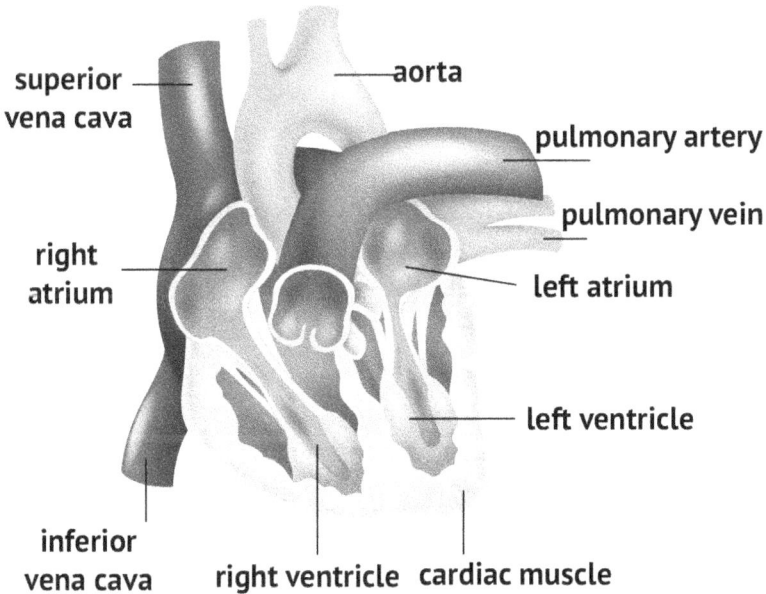

Figure 12.1. Various parts of the human heart, including the four chambers.

Blood from the right atrium (near the right arm) is passed down into the right ventricle. When this ventricle contracts, blood is then pumped through the pulmonary artery to the lungs. From the lungs and pulmonary veins, the left atrium receives freshly oxygenated blood that is passed down into the powerful left ventricle. When the left ventricle contracts, blood travels up through the aorta and to the rest of the body.

Each heartbeat is called a cardiac cycle. A heartbeat consists of two periods called diastole and systole. The diastole is where the heart relaxes and expands to refill with blood from the two circuits. The systole is where the heart has a period of robust contraction, pumping blood to the two circuits. When you feel a heartbeat, this is the heart contracting.

In order to generate the required squeezing force, the heart has powerful muscles. In fact, most of the tissue of the heart is cardiac muscle. Like so much of the human body, the heart muscles are multi-functioning, performing the actuator function at the same time as forming the chambers of the heart.

Unlike skeletal muscle, cardiac muscle exhibits rhythmic contractions and is not under voluntary control. The rhythmic contraction of this muscle is regulated by a special feature in the heart called the sinoatrial node. The cardiac muscles contract each chamber in the right order to produce the required pumping force and speed. The heart is also designed to be able to adjust its pumping rate depending on the needs of the body. If a person starts running fast, the heart beats harder and faster to increase the flow of blood and oxygen to the muscles. The volume of blood pumped by the heart per minute is called the cardiac output.

Cardiac muscle cells are organized into layers wrapped in bands around the chambers of the heart. When the individual cardiac muscle cells contract, it shortens these muscle bands, thereby decreasing chamber size. These contractions eject blood into the pulmonary (lung) and systemic (body) vessels, a pumping action coordinated through fine control from the brain and connections within the heart.

Exceptional Performance

The cardiovascular system astounds engineers and scientists. Even though brilliant engineers have used the best materials and technology, they cannot begin to match the performance of the human heart, much less the entire cardiovascular system. Here are some of the system's astounding stats and capabilities:

- There are around sixty thousand miles of blood vessels in the body.
- There are around ten billion capillaries in the body.
- The heart pumps around two thousand gallons of blood every day.
- The heart beats more than two billion times in an eighty-year lifetime.

With a few exceptions (such as bone and cartilage), cells throughout the body are no more than about two cells away from a capillary.

Of those incredible statistics, the one I find most mind-blowing is the last one. There are around thirty trillion cells in the body, and cells throughout the body are typically no more than 0.1 mm away from a blood vessel. That is staggering. It's one reason engineers are in awe of the circulatory system. Even though the blood is delivered through one artery from the heart—the aorta—that one artery branches into billions of capillaries that weave their way into every nook and cranny of the body. Such a network is beyond comprehension, let alone replication by engineers. It is ultimate engineering.

The Origin of Our Circulatory System: An Evolutionary Story in Crisis

The theory of evolution proposes that our circulatory system evolved from simple to complex over time. However, when you look at the circulatory systems of different types of animals, you do not find a gradual change in design. Instead, you see big differences that cannot be bridged by a few random changes.

The blood circulation systems in fish, reptiles, and mammals are all closed systems because the blood is contained in a recirculatory system. In the case of very small creatures, their small size makes it possible to employ a simpler, open circulatory system. For example, insects have an open system where blood flows freely throughout their bodies, lubricating tissues and transporting nutrients and waste. Insects also have a different kind of blood, called hemolymph, which does not contain red blood cells.

Closed blood circulation systems are found in most vertebrates and in some invertebrates, like earthworms. Fish have a relatively simple system with one circuit and a two-chambered heart. The deoxygenated blood goes from the tissues to the atrium and then to the ventricle where it is pumped to the gills to pick up oxygen and then sent to the tissues.

When we get to reptiles, we see the more complex and more efficient double circuit, which is more suited to life out of water, where most reptiles live. In general, reptiles have a three-chambered heart with two atria and one ventricle, which pumps blood through three arteries, one to the lungs and two to the tissues. The right atrium receives deoxygenated blood from the tissues while the left atrium receives oxygenated blood from the lungs. Since both atria open into one ventricle, the two types of blood get mixed, rendering three-chambered hearts less efficient than four-chambered human hearts at outputting oxygen. However, the power and lower level of oxygen in the blood suffices for the slower lifestyles of reptiles, and the design has the advantage of requiring less energy to operate.

So how might a blind evolutionary process build the fish circulatory system from simpler precursors, then the reptile system from the fish system, and then the mammalian system from the reptilian? Evolutionists paint a picture using very broad brushstrokes and in a way that may seem plausible initially. But when we press for details, problems quickly begin to surface and multiply.

(1) The closed blood circuit in fish is irreducibly complex

The first challenge to the evolutionary story is that even the simplest closed blood circuits, such as that of fish, are irreducibly complex, needing both a pump and a closed circuit at the same time to function. A pump on its own is no use and a circuit on its own is no use. Even a pump by itself is irreducibly complex because it needs an input, an output, and a means for pumping. And the piping on its own is irreducibly complex because it needs to form a closed circuit with connections to organs.

A further aspect of the fish circulatory system's irreducible complexity is the parallel circuitry. Think of a house's water heating system. A typical water heating system has a pump (like the heart) and pipes (like our blood vessels) in a closed circuit with several radiators (akin to organs). In addition, the radiators are often plumbed in parallel to keep pressure high, just as with the parallel plumbing of organs. A heating system often has a filter to remove dirt. In a similar way the liver acts as a filter to detoxify the blood. The house plumbing system is obviously irreducibly complex. Many parts must be present and correctly assembled for it to work at all. You cannot have half a water heating system or three-quarters of one. You need the whole system. The same is true for the fish circulatory system. For it to work at all, there are many parts that need to be present and correctly assembled from the beginning. So then, how could evolution build it one small step at a time, while maintaining function at every stage?

In 2013 some brave biologists attempted to explain the "Evolutionary Origins of the Blood Vascular System and Endothelium." Their paper said a great deal about the advantages of a closed circulatory system, but they could not say much about how such a system could arise by mindless evolutionary forces. "The ancestral vertebrate developed the means to assemble endothelial cells into tubes, to recruit pericytes and stabilize blood vessels, and to specify different subtypes of vessels, including arteries and veins," they stated. "The extent to which these processes 'borrowed' from established building plans is unclear."[2]

Their full analysis of the *how* is only slightly less fuzzy than what is found in the above passage. Such vague storytelling is typical of evolutionary accounts generally. And it isn't for lack of trying. The vagueness, rather, stems from an absolute inability to provide a detailed, causally adequate account of how the system in question could have evolved without the benefit of foresight, planning, and skillful implementation.

The statement "The ancestral vertebrate developed the means to assemble endothelial cells into tubes" is a sweeping "evolution did it" claim that gives precisely no explanation whatsoever as to how the tubes could have evolved. The suggestion "borrowed from established building plans" is a common one employed by evolutionists when faced with the challenge of irreducible complexity. To neutralize the challenge, the evolutionist explains that there is a similar system or subsystem somewhere else that was "borrowed" or co-opted for this system. But such an explanation invites at least two questions: How precisely did the borrowing occur? And how did the borrowed system arise given that it, too, is irreducibly complex? And finally, the word "unclear" is a euphemistic way of saying that investigators are at a loss as to how the system could have evolved.

It is interesting to note that the paper mentions that not many people have attempted to explain the evolutionary emergence of circulatory systems. Actually, it would be difficult to know how many people have attempted such an explanation, since biologists aren't generally in the business of publishing failed attempts to buttress evolutionary theory. Given the high praise and career rewards that await any evolutionary biologist who successfully works out a detailed and credible evolutionary pathway for some complex system (never before accomplished), it's entirely possible that many have begun the attempt but then, each in his or her turn, threw over the attempt as hopeless.

(2) The double circuit in reptiles is irreducibly complex

When we move from fish to reptiles, we find that the double circuit of the reptilian circulatory system has an added layer of irreducible complexity. Even if you could start with a single-circuit system, it is impossible to jump to a double-circuit system in a series of plausibly modest steps. One reason is that you need a change in design of the heart (e.g., the number of chambers) as well as an entire new circuit with the required connections to the heart.

(3) Our four-chamber heart is irreducibly complex

The change from the three-chambered reptilian heart to the four-chambered mammalian heart is a smaller transition than the two previous cases, but it still requires major changes to the plumbing of the circulatory system. Interestingly, the reptiles of the order Crocodilia, which includes both crocodiles and alligators, are unusual, having a four-chambered heart. Of course, the same challenge of irreducible complexity applies to crocodile hearts. Incidentally, the functional reason crocodilians have a four-chambered heart is that it is an optimal solution for them given their higher power requirements, being generally larger and more powerful than other extant reptiles.

According to evolutionary theory, the four-chambered heart has evolved independently in mammals, birds, and crocodilian reptiles. So the evolutionist needs to rely on three evolutionary miracles of irreducibly complex engineering.

The above analysis of various evolutionary proposals for getting from two chambers to three, and from three to four, barely hints at the many severe problems facing the evolutionary account. For a deeper but still accessible dive, see the article entitled "The Vertebrate Animal Heart: Unevolvable, Whether Primitive or Complex."[3]

Answering Claims of Bad Design

Despite medical experts' high praise for the design of the human heart and circulatory system, evolutionist Richard Dawkins claims the heart is poorly designed. "I think it would be an instructive exercise to ask an engineer to draw an improved version of, say, the arteries leaving the heart," he writes. "I imagine the result would be something like the exhaust manifold of a car, with a neat line of pipes coming off in orderly array, instead of the haphazard mess that we actually see when we open a real chest."[4]

Why does Dawkins say the heart is poorly designed when experts in cardiology say it is a brilliant design? He shows no evidence of having carefully studied the heart, and yet he expresses confidence that it is a bad design. The answer would seem to be that Dawkins, rather than carefully following the evidence, is again being led along by his commitment to evolutionary theory.

A look at the evidence suggests a very different picture of the human heart and the circulatory system than the one Dawkins draws, as seen in the aforementioned judgment of Sian Harding, past president of the European Section of the International Society for Heart Research, that top engineers have failed to match the ingenious engineering of the heart, and this despite hundreds of millions of research dollars and decades of assiduous effort.

Dawkins makes a gross error in assuming that the evenly spaced piping in car engines is superior to the unevenly spaced arteries and veins of the heart. Indeed, one of the weaknesses of human engineering is our tendency to use even spacing. Even spacing rarely corresponds with optimal design, but engineers often opt for it anyway because it is cheaper and easier to produce.

Evolutionist Alice Roberts also believes that the heart and circulatory system are a poor design. She has said that there should be more links connecting the two coronary arteries (which supply blood to the heart), as with dog hearts, so that if one artery gets blocked, the links would allow better blood flow to all parts of the heart.[5] But in their natural environment dogs (and their wolf ancestors) have a lifestyle with great bursts of power between long periods of rest, a mode of existence that calls for a different design of heart and blood vessels. To that end, dogs, for example, have up to twice as many pulmonary veins (veins from the lungs) to help produce those bursts of power. In contrast, humans are designed to be active throughout the day and so have a different heart design, one that is optimized for that lifestyle.

60,000 Miles of Evidence

Yes, things do go wrong with the circulatory system, such as blocked arteries, heart disease, hypertension, and varicose veins. However, these are generally caused by lifestyle choices, aging, or genetic problems. For a normal, healthy person, the circulatory system is a masterpiece of design, pumping tens of millions of gallons of blood in the course of a lifetime, with sufficient pressure to transport blood to every part of the body, delivering essential chemicals to the body's roughly 30 trillion cells, and removing waste. It's also a smart system, one that can distribute the right amount of blood and heat to exactly the right areas. All this and more is why cardiologists like Sian Harding, along with other medical specialists with expertise in various aspects of the circulatory system, see in the human circulatory system ultimate engineering.

The claim that the human heart is poorly designed is another example of how evolutionists allow themselves to be guided by a prior commitment to the Darwinian paradigm rather than following the science. The evidence itself points not to a blind evolutionary process but to an ingenious designer whose skill our best artificial-heart engineers can as yet only dream of matching.

13. The Digestive System

Two Views of the Digestive System

"Digestion of food in the human digestive system," comments scientist and food engineer Ilkay Sensoy, "is a complex combination of versatile and multiple-scale physicochemical processes" involving "disintegration to suitable forms, absorption of the basic units, transportation to related organs, and purging the remaining waste."[1] Here Sensoy gives us a glimpse into the immense complexity and sophistication of the human digestive tract. The system has the complex and challenging task of transporting food while breaking it down into nutrients such as glucose and absorbing them into the bloodstream. To pull this off, the digestive system employs a set of organs and muscles primarily coordinated by the enteric nervous system.

Engineers, and food engineers in particular, are in awe of the fine-tuning and efficiency of the digestive tract. Among the tasks of food engineers is converting corn into corn syrup and corn oil. These processes require much complex machinery and energy despite the help of solvents and microorganisms. And yet the body can turn corn and a range of other foods into nutrients in a small space with minimal energy.

The digestive system has to work seven days a week for a lifetime. Muscles move food along thirty feet of tubes using a special system of muscle-driven waves. Muscles also churn the food in the stomach and mix the food in the small intestine. The biomechanics and layout of the process are a masterpiece of design.

But evolutionists from Darwin to Hafer have failed to acknowledge the brilliant design of the digestive system and stick to evolutionary theory's mantra of bad design. As we shall see, the claim of bad design is not supported by science.

Digestion: It's Complicated

Digestion is a fine-tuned process involving the coordination of at least twelve elements, each requiring a complex system in its own right, and most of the activity occurring without our conscious control. This system of systems performs the following tasks:

- Generates a desire for food through feelings of hunger.
- Produces saliva (containing specialized enzymes) from glands to help digestion.
- Breaks down food mechanically in the mouth into balls of food called boluses.
- Breaks down food mechanically using stomach muscles.
- Breaks down food chemically in the stomach and small intestine using specialized enzymes.
- Produces bile in the liver and stores it in the gallbladder to help digest fats.
- Propels food through the digestive tract using muscles.
- Houses the microbiome—consisting of trillions of microbes, fungi, viruses, and archaea—which plays an essential role in digestion.
- Contains 70 percent of the immune system.
- Absorbs nutrients through digestive tract walls.
- Removes waste from blood using the liver.
- Stores waste in the bowel.
- Allows waste to leave the body using sphincter muscles.
- Controls the digestive system through the enteric nervous system (ENS).

Food usually takes two to five hours to empty from the stomach, two to six hours to pass through the small intestine, and ten to thirty hours to move through the large intestine—over forty hours total.

The uninitiated might picture digestion as primarily a matter of grinding up the food mechanically in the mouth and stomach, with saliva and stomach juices brought in for some vaguely understood lubrication duties. But chemically breaking down food is a complex process.

In foods like bread and potatoes, a specialized enzyme in the saliva and pancreatic juice, called amylase, breaks the starch into sugar molecules called maltose. Then another enzyme, called maltase, in the lining of the small intestine, breaks the maltose into two glucose molecules.

Fats are broken down into small droplets by the bile produced in the liver and sent to the small intestine. This helps other specialized enzymes called lipases, which are sent to the small intestine from the pancreas to break down the fats into fatty acids. To further help with digestion, bile is stored and concentrated in the gall bladder to be sent to the small intestine soon after a meal with fat in it is consumed.

Proteins are broken down into amino acids by a suite of specialized enzymes called proteases, which are produced in the pancreas and sent to the small intestine. The stomach has its own special protease called pepsin.

The above paragraphs are an extremely simplified sketch of an enormously more complicated process that ensures that the body's various enzymes and other chemicals digest what should be digested and do not digest what shouldn't be digested. For instance, proteases, coming from the pancreas to the small intestine, are kept inactive to avoid self-digestion, and through a precisely orchestrated signaling, delivery, and activation system, are put into action only when and where needed.

The digestive system is so efficient that almost all proteins, carbs, and fats are absorbed. Fiber is the only part of food that does not get digested. It's important in the diet, however, as it helps move food through the digestive tract. Feces contain mainly water, fiber, bile, and bacteria. Many different types of bacteria live in the intestines, and these help the body break down carbohydrates. Gut bacteria also provide critical nutrients, like vitamin K.

What I find especially impressive is so many processes acting at different scales, from macroscales (peristaltic muscles) to microscales (nutrient absorption). This, and the masterful blending of mechanical and chemical processes with the help of the nervous system.

Anatomy of the Digestive Tract

The digestive tract is an open-ended tube some thirty feet long from mouth to anus. Its main parts are the pharynx, esophagus, stomach, small intestine, and large intestine. We looked at swallowing in Chapter 6, so here we will focus on the stomach, small intestine, and large intestine. The small intestine is upwards of twenty-five feet long. The large intestine (also known as the colon) is about five feet long. To fit inside the body, the digestive tract is folded many times.

The inside of the stomach has a mucous membrane that enables transfer of fluids into and out of the stomach. The outside of the stomach has a complex arrangement of three muscle layers that enable the stomach to be squeezed in different directions. The muscles, as we saw, work to break down the food mechanically. They also move the pre-digested food towards the small intestine.

The small intestine has three segments—the duodenum, jejunum, and ileum—with each segment specialized for a different aspect of digestion. It's around one inch in diameter, and its inner surface is highly undulating with villi and microvilli, significantly increasing the surface area for more efficient absorption. The small intestine's surface area works out to around 2,000 square feet, almost the area of a tennis court. The huge workload of the villi requires their replacement about every five days. This is possible through stem cells found in the intestinal glands. Bacterial density is low in the stomach and proximal small intestine but rises progressively toward the large intestine, driven by decreasing levels of acid, bile, and oxygen.

The small intestine also has four layers. (See Figure 13.1.) From external to internal they are serosa, muscularis propria, submucosa, and mucosa. The outer layer provides support and contains blood vessels and nerves. The muscularis propria is the muscle system that contracts and relaxes during digestion. The submucosa supports the surrounding layers and provides elasticity, while the mucosa acts as a physical barrier and aids absorption of nutrients.

What remains of the food moves from the small intestine to the large intestine. The large intestine is divided into the caecum, colon, and rectum. This region is home to a dense, diverse garden of microbes, viruses, fungi, and archaea. Collectively these create a chemical factory for metabolizing resistant starch, a type of fiber, into short-chain fatty acids, the primary fuel of colonocytes, which are special epithelial cells that line the inside of the large intestine and are crucial in the latter stages of digestion and absorption. Over the past fifteen years, research has revealed that the colonic microbiome is essential to human health, driving energy harvest, gut barrier integrity, immune tolerance (preventing autoimmunity), pathogen resistance, metabolic regulation (preventing obesity), and neuropsychiatric stability (reducing mood disorders).

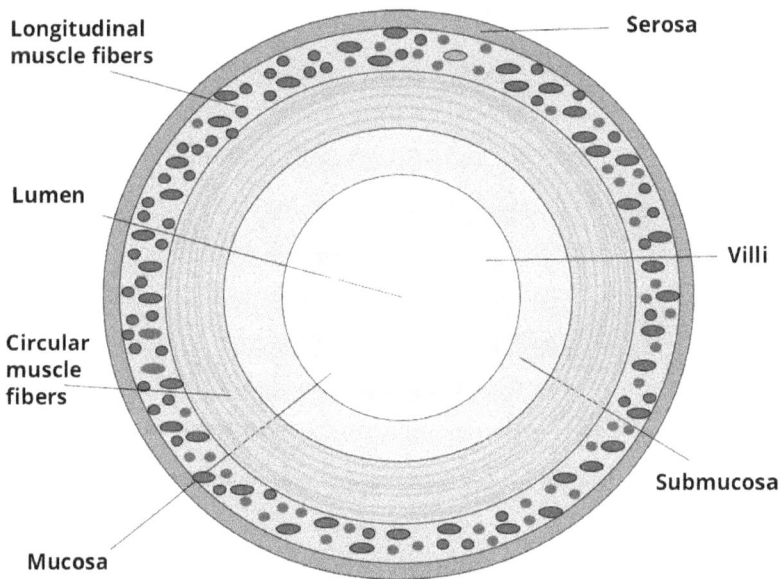

Figure 13.1. Cross section of the small intestine tube.

Biomechanics of Food Transport

When food enters the stomach, it is mixed with gastric juice. Strong muscular contractions in the stomach wall are triggered and work in tandem with the gastric juices to reduce the food to chyme, a thick milky material. The sphincter muscle at the lower end of the stomach slowly releases chyme into the duodenum.

There are two main types of muscle movements in the digestive tract: peristalsis and segmentation. Peristalsis occurs in every part of the tract and commences when nerves trigger two sets of muscles in the gut wall to initiate a series of wave-like contractions that push the bolus forward. In segmentation, circular muscles in the intestines contract to move food back and forth, like a washing machine. This churning mixes the food with gastric juices, helping break it down into smaller pieces. Whereas peristalsis moves food along, segmentation slows the food's progress through the digestive tract, ensuring more thorough digestion of nutrients.

PERISTALSIS MOTION

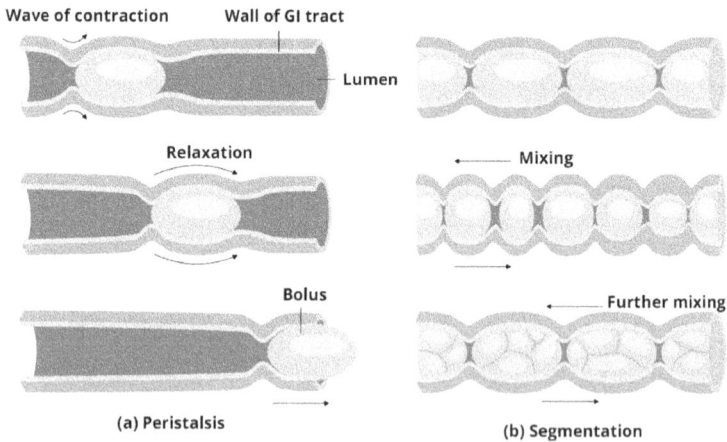

Figure 13.2. Peristalsis and segmentation.

Humans, like many mammals, have a vomiting reflex whereby the wave-like muscle contractions of peristalsis shift into reverse. This is known as reverse peristalsis or antiperistalsis. Vomiting is unpleasant but it serves to expel pathogens or undigestible items, and to remove excessive content that otherwise might overload and damage the digestive tract. Reverse peristalsis can carry food from your small intestine all the way back through your stomach, esophagus, and mouth. At times, less extreme measures suffice. If there is a blockage in the intestine, reverse peristalsis may move food backward only a little way and then the body tries again to digest the food, often successfully.

The Enteric Nervous System: A Second Brain

The complicated processes in the digestive system are regulated by an autonomic circuit of neurons called the enteric nervous system (ENS). The ENS is sometimes referred to as a "second brain" because of the complexity of the task. The ENS, itself stimulated and modulated by the parasympathetic nervous system, controls gut muscle movements, constantly fine-tuning the digestive system to coordinate processes in the most efficient way.

The ENS takes into account what is being eaten and how much, and instructs processes accordingly. Its job involves coordinating the muscles of the entire digestive system. When food leaves the stomach, for instance, the muscles of the small intestine must start to work immediately. And the muscles must be controlled to move food along the digestive tract at just the right speed. If it is too fast, the food cannot be digested; if it is too slow, digestion is inefficient. The ENS also coordinates the release of enzymes that break down food and the flow of blood in the digestive tract that helps with nutrient absorption. It also maintains the balance of gut microbiota. The ENS coordinates all this.

Building the Digestive Tract

What most impresses me about the digestive tract is how it gets constructed during embryological development. Growing the digestive tract in the first few weeks of life in the womb is a very challenging design task because the intestines and stomach grow much faster than the rest of the body, due to the complexity of the digestive tract.

Around the fifth week of life, the midgut has lengthened much more compared to the abdomen, and this means that the intestines must have a more compact shape at this stage compared to their final configuration. The result is that the digestive tract has to undergo large shape changes as the baby grows in the womb. Research has revealed that the digestive tract also goes through some remarkably elaborate movements by the end of pregnancy. In particular, the stomach goes through two rotations in different directions, and the embryonic midgut (which becomes most of the large and small intestines) has to go through three rotations to get to the required shape.

As the journal *Nature* reports, not much is known about how the gut makes these rotations. "A protein found on the embryo's left side was long thought to be the major regulator of left–right asymmetry in organ development," but based on animal experiments, researchers at Cornell University proposed that the connective-tissue molecule hyaluronan, situated on the right side of the gut, "plays an unexpected part in controlling intestinal asymmetry." The researchers found that, in mouse and chick embryos, "hyaluronan becomes decorated with amino-acid chains, but only on the right side of the gut. An accumulation of modified hyaluronan causes the gut's right side to expand significantly, tilting the organ leftward and triggering rotation. Blocking this modification led to mouse embryos with abnormally twisted guts."[2]

The clever rotations of the digestive tract during embryological development remind me of the solar array deployment system I designed for the European Space Agency's METOP meteorological satellites. The solar panels had to be rotated three times in order to get them into the right shape and alignment. Achieving these three rotations required an immense amount of foresight and planning. In the same way, engineering the system that rotates the digestive tract in the embryo required an immense amount of foresight and planning.

A Vestigial Argument

One part of the digestive system that wasn't mentioned in our trip through the digestive tract is the appendix. Evolutionists have claimed that the appendix is an evolutionary leftover and a bad design feature. In speaking of it, Hafer said she found it hard "to see why a Creator making us in His image would give us an organ that works best in rabbits."[3] Her remark follows a long tradition stretching back to Darwin himself. "Not only is it useless," commented Darwin, "but it is sometimes the cause of death."[4]

Darwinists have been repeating that claim ever since, and holding the human appendix up as evidence of the blind evolutionary process's penchant for bungling the creative process. They have been repeating this claim despite the fact that at least as early as 1900 it had been shown that the appendix is far from useless.[5] Certainly one can make do without an appendix. This is true of certain other body parts as well. But this is because the body has built-in redundancies, itself a hallmark of well-engineered systems. At the same time, there are now multiple studies which together show that the appendix has a dual function: fighting infections and storing good bacteria.[6] One could describe the human appendix as a built-in probiotic. This small, finger-shaped pouch, located at the junction of the small and large intestine, harbors beneficial gut bacteria that are protected from being flushed out during digestive disturbances such as diarrhea or anti-biotic use.[7] These microbes are subsequently reseeded into the colon to restore a healthy microbiome. In children, the appendix also plays a crucial role in the development of gut-associated lymphoid tissue (GALT), thereby supporting early immune maturation.[8] Physiologist Loren G. Martin summarized the matter in *Scientific American*: "For years, the appendix was credited with very little physiological function. We now know, however, that the appendix serves an important role in the fetus and in young adults."[9]

Claims that the appendix is a poor design—an evolutionary leftover—reflect a lack of knowledge as well as a lack of curiosity on the part of evolutionists, a lack encouraged by their tendency to view as evolutionary junk any biological system for which they cannot readily ascertain a function. To argue that the digestive tract wasn't intelligently designed because the appendix appeared at first glance to be useless is akin to an uninformed car owner concluding that his Lexus automobile has an engineering defect because, as far he can tell, the mysterious differential lock button doesn't do anything useful. This misconception has influenced medical training and culture, leading many physicians to view the appendix as biologically unimportant. Recent research challenges this perspective, demonstrating that appendectomies are associated with a spectrum of subtle but adverse health outcomes, including gut dysbiosis and increased susceptibility to recurrent infections.[10]

The appendix is in good company, by the way. Evolutionists have branded various other organs and parts of the human body, including the tonsils, the coccyx, and "junk DNA," as largely useless evolutionary leftovers. But as with the appendix, these too have been found to serve important purposes.[11]

Digesting the Evidence

Given everything the human digestive system is able to handle—everything from vegetarianism to Inuit diets heavy on sea mammal meat and blubber and all but devoid of plants—it is surprising that evolutionists feel brash enough to criticize the design.

Digestion is an immensely complex process generally carried out with great efficiency even as the person is barely aware of the hard work going on inside the body. The main conscious feedback is the pleasure involved in eating, the feelings of hunger and fullness, and the experience of indigestion, warning one to back off certain dietary choices that may be overtaxing the system.

As with every part of the body, things can go wrong. Problems include acid reflux, irritable bowel syndrome, hemorrhoids, and colitis. In some cases, the problems stem from lifestyle factors, such as poor diet and lack of exercise. In other cases, problems are caused by genetic decay or old age. But none of these problems can be attributed to design flaws in the canonical human digestive system, at least if well-established criteria for assessing design quality are employed. When those criteria are employed, what becomes plain is that the digestive tract is an incredible feat of engineering, beyond anything humans have been able to engineer.

14. Muscles and Tendons

Two Views of Muscles and Tendons

That our muscles and tendons show evidence of next-level engineering prowess is not a secret. George Székely, an expert in muscle research, titled a paper in the *Journal of Brain Sciences* "A Perfect Design: The Multifunctional Muscle."[1] And *Britannica*, which can generally be counted on to convey the conventional view on a subject, describes tendons as "remarkably strong, having one of the highest tensile strengths found among soft tissues. Their great strength, which is necessary for withstanding the stresses generated by muscular contraction, is attributed to the hierarchical structure, parallel orientation, and tissue composition of tendon fibres."[2]

I have had a research team develop artificial muscle in the lab at Bristol University. In 2012 we published an award-winning paper on artificial muscle in the journal *Smart Materials and Structures*, a top journal in the field.[3] My research team designed an artificial muscle using layers of dielectric elastomers with graphite for electrodes and using a slender metal plate to fine-tune the direction of movement. We applied large voltages to get large displacements similar to that of muscles. One of the main lessons we learned was that even if we used the best materials available, we could not approach the performance of animal muscle in terms of power-to-weight ratio.

Perhaps even more impressive, animal muscle is much more than an actuator. Here are five functions of muscle:

1. Serves as an actuator for movement and force.
2. Pads/protects most of the body.

3. Aids in thermoregulation.
4. Gives an efficient shape to the body.
5. Gives an elegant shape to the body.

Thus, every muscle in the body is not only well sized and precisely placed for actuation but is also well suited for these other functions. Engineers are decades away from getting this level of multifunctionality from artificial muscles.

Despite the brilliance of natural muscle, evolutionist Nathan Lents tries to pick fault with its design. As before, I fully agree with Lents that evolutionary theory predicts bad design. However, as we shall see, the scientific evidence strongly points to animal muscle being supremely well designed.

An Intricate Hierarchical Structure

Muscle has an intricate hierarchical structure. Each muscle includes many individual motor units independently controlled with their own nerve pathways. Most muscles contain between 100 and 1,000 motor units. There are over 100,000 muscle motor units in the whole body. That is an astounding number of individually controlled actuators and vastly more than in any robot design. Each individual motor unit consists of many muscle fibers, and all together the body contains millions of individual muscle fibers.

Muscle's hierarchical structure allows for fine-tuning of design. If a muscle is required only for coarse movements, the number of muscle fibers for a given number of motor units will be relatively high. The innervation ratio represents the number of muscle fibers controlled (innervated) by a single motor unit. For example, the gastrocnemius muscle of the calf has innervation ratios as high as 2000:1. In contrast, finger muscles have innervation ratios as low as 10:1, which afford fine control.

Unlike a robot, muscles do not require bulky batteries. Instead, blood is pumped into the muscle, supplying oxygen and glucose for energy. This makes the human body much more compact.

Structure of Skeletal Muscle

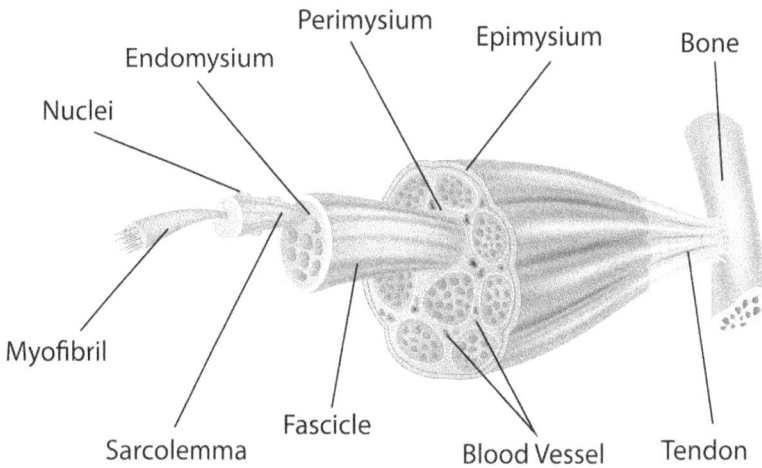

Figure 14.1. The hierarchical structure of muscle.

Hundreds of Muscles in Symphonic Harmony

There are over 600 muscles in the human body, each well designed for its function. One of the most common types is skeletal joint muscle. These muscles are used to move or prevent movement in the 360 joints in the body. There are muscles in the diaphragm of the chest, which control breathing. There is the cardiac muscle for pumping blood. There are numerous facial muscles for making nuanced facial expressions. Many of our muscles we can control voluntarily, while others are controlled by the autonomic nervous system, such as the peristaltic muscles in the intestines that move food through the digestive tract.

We tend to think of muscle as an actuator that only causes movement. However, muscle is also used to *stop* movement. Muscles either apply force as they move (isotonic contraction) or resist force while stationary (isometric contraction). Examples of where muscles change length are movements of the joints in the leg while walking or running. Examples of where muscles remain stationary while resisting force include holding a weight with a stationary hand and maintaining a posture while standing.

Powerful and Precise Actuators

The largest muscle in the body by volume is the gluteus maximus (located in the buttocks), followed closely by the thigh's powerful quadriceps femoris (actually four muscles), which extends the knee joint and flexes the hip joint. The body's longest muscle is the sartorius, which runs down the length of the thigh and complements the quadriceps' work of moving the hip and knee joints.

Human muscle also has a very high power-density, up to 400 watts per kg. This means that two quadriceps weighing 3kg total will have up to 1,200 watts of power. This explains why Olympic cyclists can maintain super high speeds in track cycling. Despite intense research efforts, engineers cannot match this power density in artificial muscle.

To illustrate the superiority of human muscle, arm wrestling competitions have been staged between humans and robots. In these competitions, robots are restricted to having artificial arms that are no bigger than human arms. In addition, the robot arms are restricted to using flexible muscle-like actuators. One competition was held March 7, 2005, at a conference in San Diego on Electroactive Polymer Actuators and Devices (EAPAD). Three leading robot teams entered robots and competed against a female student from the San Diego school district. The student easily beat all the robot arms.

One of the smallest muscles in the body is the stapedius muscle in the middle ear. It is around 0.2 inches (5 mm) long and weighs just a couple of milligrams. The gluteus maximus is hundreds of thousands of times larger than the stapedius muscle. This demonstrates the incredible versatility of muscle design.

Fine Control

It is easy to take for granted our ability to hold delicate objects like eggs and insects. However, to do this requires advanced engineering solutions that engineers struggle to replicate in robots. Holding a fresh egg requires maintaining just the right level of force and friction: too little and the egg will be dropped; too much and the egg will crack. To maintain the right level of force requires fine control of several muscles, along with feedback from sensors in the hand about what force is being applied.

Muscles are notable for their ability to produce smooth motion. When a signal is sent to a group of motor neurons to perform a movement, they're not all recruited at the same time. Instead, the brain gradually activates more and more individual motor units within individual muscles in order to create a smooth increase in force.

The control of muscles is designed so that smaller motor units are recruited first in order to maximize the smoothness of force take-up. Each muscle also has many position sensors (proprioceptors) to aid accurate control. Interestingly, if you move your arm and hand joints with your eyes closed, your brain knows the position of your arms and fingers thanks to the position sensors in your limbs. This is why it is possible to move around in the dark and why blind people can be just as skillful with their hands as people who can see.

One reason for the fine control of muscles is that they often work in antagonistic pairs, such as the biceps and triceps. Since muscle can only contract, joints generally have a set of opposing muscles that can pull the joint in opposing directions. The brain controls these antagonistic muscles precisely so that as one muscle contracts, the other relaxes. The timing is so precise that there is no noticeable slack, even for changes in direction.

Robot engineers go to great trouble to create smooth motion and efficient actuation. Yet the best artificial muscle actuators do not begin to compare with the performance of animal muscle in this regard. Robots really struggle to perform activities such as playing the piano, knitting, or even holding an egg.

High Efficiency

Animal muscle is far more efficient and practical than any human-made actuator. Muscle cells perform a process called respiration, which involves breaking down glucose and other respiratory substrates to make the energy-carrying molecule called ATP. ATP is like a charged battery. When ATP is converted to ADP, energy is released to contract the muscle and produce force. The body is continuously converting ADP back into ATP ready for energy release.

This is a great design solution because sugar and oxygen are readily available and sustainable types of fuel. In addition, the blood circulatory system continuously provides glucose (from the small intestine, liver, and kidneys) and oxygen (from the lungs).

The conversion of glucose and oxygen into energy is extremely efficient for most muscle usage, which involves low and moderate power levels. For these power levels there is aerobic respiration, where glucose and oxygen are converted into ATP. For activities requiring a high power level, there is anaerobic respiration, where just glucose is converted into ATP.

One feature that increases the efficiency of both forms of metabolism is that the heat produced in the chemical reactions can be used to warm the body through heat transfer to the blood. (More on this below.)

Glucose is stored (in a form called glycogen) in muscles and the liver. During exercise, as glucose stores decrease in the muscles, reserves are transported from the liver to the muscles. After long periods of exercise, the muscles fatigue and ache. In order to get extra glucose into the body quickly during sports and other strenuous activities that persist over many minutes or hours, people can consume natural sources of glucose, such as honey or grapes, or opt for energy gels and energy drinks.

Engineers would love to be able to make a robot using such a sustainable form of energy. Instead, engineers have to put heavy batteries on their robots that add significant weight and bulk. Of course, a gasoline engine would have a high power-to-weight ratio, but this solution is rarely desirable due to issues such as vibrations, noise, safety, and pollution.

The Padding Function

Now let's consider other important functions that muscles fulfill, starting with the padding and protection functions. Many people are unaware that muscles provide important protective barriers for bones and organs like the stomach, bladder, kidneys, and liver. Muscles create a cushioning barrier that can absorb impacts. Even though impacts cause bruising to the muscle, this bruising heals quite quickly due to the good blood supply to muscles.

The glute muscles in the bottom provide an important padding for sitting and also for falling. The glutes are the largest and thickest in the body, which helps make sitting comfortable. When falling over backwards, it is natural to fall on glute muscles in the bottom. These muscles help protect the spine from injury.

The hands are a good example of designer padding. The hands and fingers have areas of padding ideally positioned to provide cushioning and protection. The hand muscles are precisely shaped and located so as to form these pads. Additionally, there are fat layers added to these pads for extra cushioning. There is also a small fat pad on the palm side of each finger, which cushions the fingers when they grasp objects tightly. These cushioning pads are ideally placed not only for protection but also so as not to hinder the fine sense of touch and skill of the fingers and palms.

The Thermoregulation Function

In most climates humans live much of the time in temperatures lower than the body temperature of 98°F, so there is a need for the body to generate significant heat. The muscles play a major role in generating that heat.

Skeletal muscle makes up around 40 percent of the mass of humans and can contribute up to 85 percent of the heat of the body. Muscle is ideal for providing heating because blood is highly integrated with the muscles of the body. Every muscle has several blood vessels, so heat transfer is very effective. Exercise can maintain a warm body temperature even in fairly cold weather. When athletes are running at a brisk pace in near-freezing conditions, they produce so much heat that they can run in just shorts and t-shirts.

When the body gets very cold, it responds by shivering. This causes the muscles to go through repeated contractions, which expend energy and generate heat. It is remarkable that the "waste" heat of metabolism is so well utilized for an important function. That is a hallmark of brilliant engineering.

The Body-Shaping Function

The human body has an aesthetically pleasing shape due to the muscles, especially when the person is fit and healthy. It's so impressive that muscles can be so functional while, at the same time, contributing to an aesthetically pleasing shape. Yes, our post-industrial Western society encourages physical inactivity and an unhealthy diet, leaving many of us far from fit. And there are the inevitable effects of degradation from aging. But there is beauty in every human being. When teaching aesthetics to students I point out that modern cars are given "character lines" in the panels to create beauty. The term *character lines* comes from wrinkles in the human face. In past ages, wrinkles were considered by many a positive feature because they give character to the face.

True, elegance and beauty are partly subjective, but *subjective* shouldn't be equated with unreal; and in any case, elegance and beauty are only partly subjective. This isn't the place for an extended excursion down the rabbit hole of aesthetics. Suffice to note that there are well-established objective reasons for considering the human body both beautiful and elegant. One of the purposes of Leonardo da Vinci's Vitruvian Man sketch (Figure 14.2) was to highlight the balanced proportions of the human body. The body fits a square when the arms are outstretched and a perfect circle when the body does a star jump. The graceful curves of the body and limbs also contribute to the beauty of the human form. In the case of the female body, there are additional curves and areas of fat that make a complementary beautiful shape to the male body.

The body is not just an elegant shape but also efficient for moving—form and function harmonize. Notice how the calf muscle is concentrated towards the top of the lower leg. In fact, there is little calf muscle at the bottom part of the leg because the muscle changes completely to a narrow tendon by the bottom of the tibia. This concentration of muscle at the top of the tibia is ideal because it gives the lower leg a low moment of inertia and hence efficient movement. Moment of inertia is a property that affects how efficiently the leg moves in rotation. Whereas mass affects how quickly a body can accelerate along a straight path, moment of inertia affects how quickly a body can rotate. For a given amount of muscle mass, the way to minimize the moment of inertia is to place the muscle mass as close to the point of rotation as possible. In the case of the calf muscle, this means concentrating the calf muscle as close as possible to the knee joint. And that is exactly what is done.

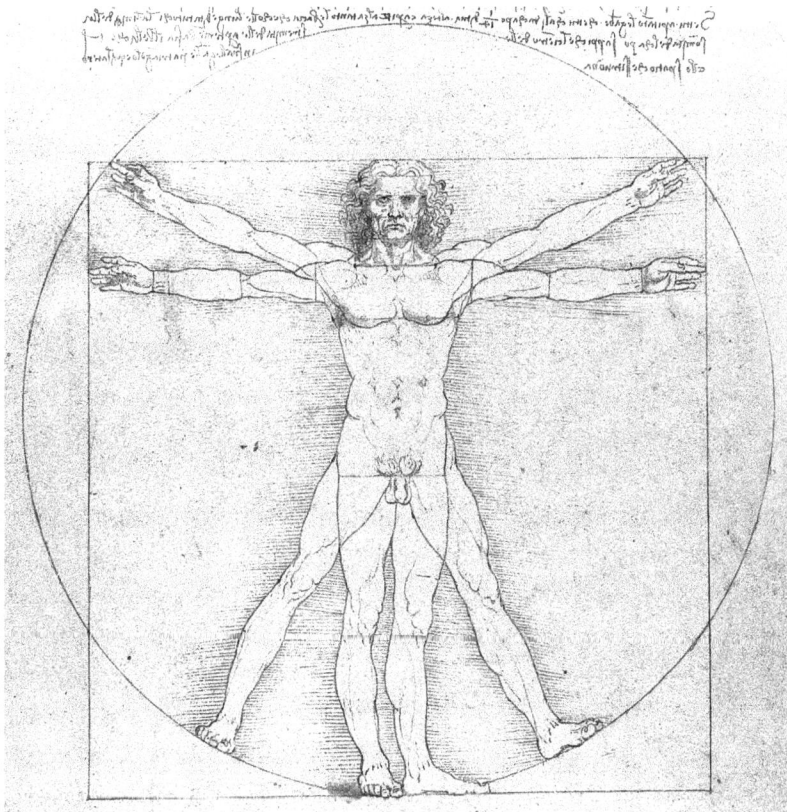

Figure 14.2. Leonardo da Vinci's Vitruvian Man.

A similar design feature is seen in the thighs, with muscles concentrated towards the hip joint. The same principle is also seen in the arm, with hand muscles concentrated towards the elbow joint. In these and other ways, form and function harmonize beautifully.

Another Lents Design-Review Failure

As eager as many evolutionists are to see and spotlight cases of bad design in biology, my sense is that complaints from them about the supposed bad design of muscle are comparatively rare, perhaps because even evolutionists with little if any engineering background can readily see how extraordinary the functionality of muscle is. When it comes to the musculoskeletal system, those evolutionists who make it their business to denigrate biological designs tend to pick on joints they consider poorly designed—joints, as we have seen, that the evolutionists poorly understand, such as the wrist and ankle joints.

Another part of the musculoskeletal system they are apt to criticize is the tendon that connects the two largest calf muscles to the heel bone, namely, the Achilles tendon, the largest tendon in the human body. Nathan Lents has this to say about it: "The poorness of this design can be summed up in the observation that the function of the entire joint rests on the actions of its most vulnerable part. A modern mechanical engineer would never design a joint with such an obvious liability."[4]

Again, I fully agree with Lents that evolutionary theory predicts bad design. The problem for Lents is that the scientific evidence points to the Achilles tendon being extremely well designed.

Lents repeats the mistake of conflating problems caused by misuse of a well-designed mechanism with problems caused by poor design. The masterfully engineered manual transmission in a high-end sports car can still be wrecked by poor maintenance, ordinary wear as the vehicle ages, or misuse of the clutch. Similarly, problems with the Achilles tendon are due not to bad design but to non-design issues such as lifestyle, overtraining, age, and disease. The most common cause of Achilles injury is overuse, especially in sports that involve explosive acceleration. These competitive activities encourage the more spirited participants to test the limits of the human body, and some athletes and their coaches are all too good at finding those limits.

Also telling: Certain demographic segments are much more prone to this injury than are others—for example, male "weekend warriors" in their forties who haven't accepted that they aren't twenty anymore.

I find it incredible that Lents would claim that a modern mechanical engineer would never design such a mechanism as the Achilles. Modern engineers regularly design systems just like the Achilles, because the tendon is a pulley-cable system, which is an important component of many modern mechanical systems.

I have often designed pulley-cable designs like the Achilles tendon for deploying spacecraft solar panels. In fact, I have won national prizes for spacecraft deployment systems that included cable-pulley systems. If Lents had consulted with mechanical engineers, he would not make such a wild claim about the Achilles tendon.

The Achilles tendon is the perfect design for the foot because its location maximizes the cross-sectional area of the tendon connection, thus maximizing the load capacity. During running, the tendon experiences loads of nearly eight times body weight.[5] For a large male, that represents a load of nearly one ton. The Achilles tendon can cope with this great load requirement because it is such a good design. As usual, Lents offers no better solution. He seems to be unaware of engineering etiquette: If you cannot think of a better solution, don't badmouth the design.

600+ Strong Arguments for Design

As for that most prominent part of our musculoskeletal system, muscle, George Székely is right to call it a perfect design. Its actuation abilities alone are astounding, and yet muscle does so much more than function as an actuator. Muscle also serves as a protective barrier and a thermal component, and it lends the bodies of humans and other animals an aesthetically pleasing shape. That muscle fulfills all these functions is no small feat of design prowess. Having run research projects to build artificial muscle, I know from firsthand experience that muscle represents ultimate engineering.

15. The Nervous System

Two Views of the Nervous System

I have a deep appreciation for the design of the nervous system, having led the design of the wiring subsystem for the solar array on the world's largest Earth observation satellite, Envisat. The system wiring required immense planning and foresight. I had to route more than four hundred wires over a combined distance of more than three miles. There were many occasions when I had to ask permission to put a hole through a critical structure in order to get the wires to their destinations. These and other issues impacting two or more design teams called for delicate compromise, collaboration, and planning. All of us who worked on Envisat's wiring systems are justly proud of the work we contributed to that advanced engineering project. But compared to the human nervous system, our work was like toddlers playing with shoelaces.

That's a whimsical way of putting it, but it's no exaggeration. The human nervous system is a vastly more sophisticated design than any wiring system engineers have ever devised, possessing a breathtaking level of detail and precision. Design and planning are written all over it. Like other engineers, I am in awe of the human nervous system.

Incredibly, despite the sheer brilliance of the human nervous system, evolutionists such as Jerry Coyne, Richard Dawkins, and Nathan Lents have tried to find fault with some of the design details. But as we shall see, their criticism is a reflection of their lack of knowledge of design principles and not of anything lacking in the nervous system's design quality.

The Human Brain: The Ultimate Supercomputer

In a book from MIT Press, researchers Peter Sterling and Simon Laughlin ask, "How can the brain be far smarter than a supercomputer yet consume 100,000-fold less space and energy?"[1] This illustrates the enormous gulf between biology and human technology. Biology is not just a little better than our technology. It is on a completely different level. All the evidence points to the human brain being not just vastly superior to our best supercomputers but at the limit of what is physically possible.

Sterling and Laughlin gush over the performance of the human brain, describing it as far superior to a supercomputer and detailing a series of design principles that allow for the brain's vastly superior information processing. These principles include: compute directly with analog primitives; compute with chemistry; combine analog processing and pulsatile processing; code sparsely; send only the information needed for a given task; transmit information at the lowest workable rate; make neural components irreducibly small; and minimize wire.[2]

That the human brain is more energy efficient than a supercomputer is probably not surprising to most of us. Supercomputers are energy-sucking behemoths. But is the human brain really also far smarter? For those unfamiliar with the fields of neuroscience and computer science, the claim of the brain's superiority to supercomputers may seem implausible if one focuses only on news stories of computers, like Deep Blue, beating chess geniuses. Undoubtedly, computers excel at executing algorithms programmed by a team of software programmers and chess experts. But Deep Blue does only what it's programmed to do. Unlike humans, computers are not creative. As Sterling and Laughlin put it rather harshly, "Despite Deep Blue's excellence at chess, it... is stupid, the electronic equivalent of an idiot savant."[3]

In 2003, Selmer Bringsjord, Paul Bello, and David Ferrucci devised a way to assess computer creativity. They called it the Lovelace test.[4] They said a computer program passes the Lovelace test if it does something that can't be explained by the programmer or an expert in computer code.

Computer engineer Robert J. Marks explains: "There have been many cases where a machine looked as if it were creative, but on closer inspection, the appearance of creative content fades.... all creativity comes from the programmer. AI reproducing the sort of creative flashes of Gauss, Tesla, or Philipp without first being given the answers in a batch of solutions is not possible."[5]

Computers can only spit out the results of what has been programmed into them. The amazing things we see them do are signs not of their own genius or creativity, but of human genius and creativity. We can innovate; computers cannot.[6]

The human mind can also disambiguate. It has common sense and understanding. Computers can't and don't.[7]

According to a straightforward application of evolutionary theory, with its dependence on random mutations and a selection mechanism with zero foresight, our brains should have a kludgy design, and our most advanced computers could be expected to readily surpass them, since engineers are not constrained to a blind evolutionary process. And yet the opposite is true. The brain outperforms our best computers, a fact most elegantly underscored by the reality that humans invent computers, not the other way round—science fiction fantasies notwithstanding.

Wired from Head to Toe and Everywhere in Between

Of course, the brain does not work in isolation but is part of a nervous system that permeates every part of the human body. Figure 15.1 provides a simplified overview of the human nervous system. It is impossible to do justice to the nervous system through diagrams, because of the sheer magnitude of the design details.

One challenge with any complex mechanical system is to get wiring to every motor and sensor. In the case of the human body, this task is unimaginably great. The human body has nerve pathways from the brain to every cubic millimeter of the body. To give a sense of how fine-grained this is, a cubic millimeter is smaller than a small grain of rice; and it takes a thousand cubic millimeters to make just one cubic centimeter. For the nervous system to be so finely branched that it can reach every cubic millimeter of the body is astounding.

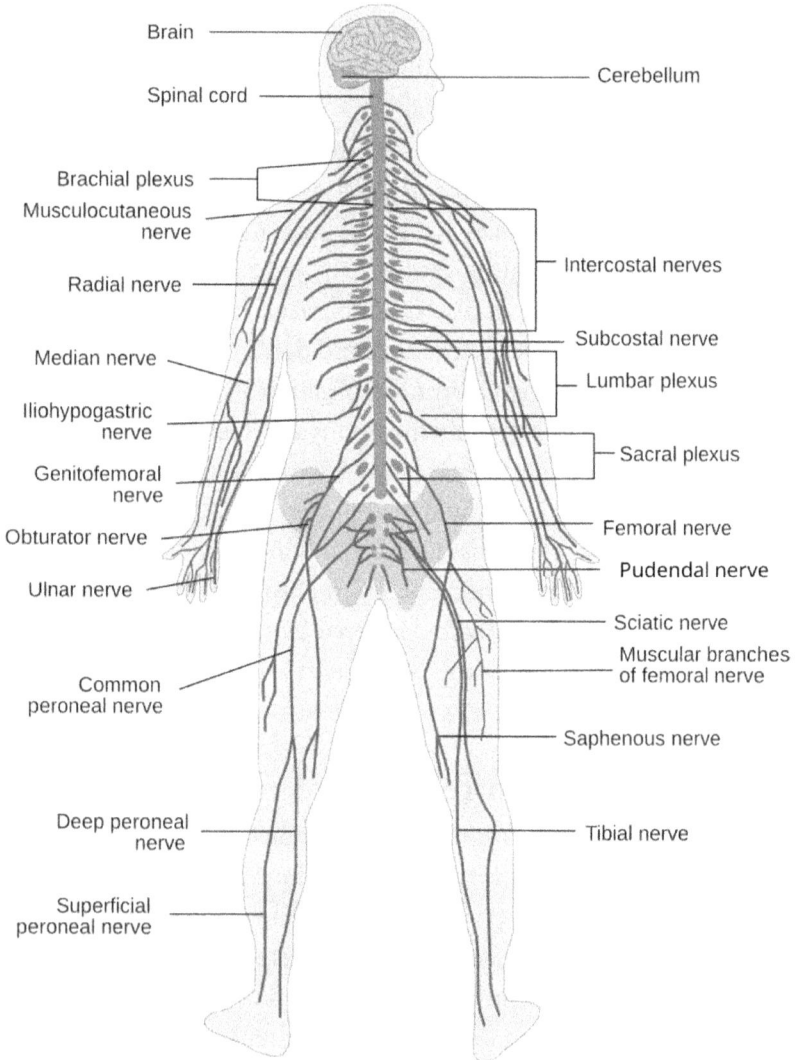

Figure 15.1. The nervous system of the human body.

How is this accomplished? The spinal cord has thirty-one pairs of main nerve branches, and these branch to smaller and smaller branches until tiny nerve pathways reach every part of the body. Nerve pathways allow messages to travel in two different directions, to and from the brain. There are many thousands of motor signals that travel from the brain to the different muscle motor units throughout the body, and there are many millions of sensory signals that travel from the body to the brain. The brain and spinal cord are the central nervous system. The nerves that branch off from the spinal cord to the rest of the body are the peripheral nervous system.

The nervous system is crucial to our getting on in the world. It's essential to, among other things, our sense of touch. If you're pushing your way through undergrowth in a forest and inadvertently press into a thorn, you'll feel an immediate shot of pain and likely a reflexive desire to pull back. This rapid warning system helps you avoid driving the thorn deeper or blundering into other nearby thorns. Importantly, this experience will occur no matter where the thorn pricks your body, because every cubic millimeter of your skin has sensors connected to nerve pathways to the brain. This comprehensive system of nerve pathways is also responsible for our delicate sense of touch. In just one square millimeter of a fingertip, there can be more than twenty nerve pathways.

Millions of nerve pathways weave their way from the spinal cord into every muscle unit, every tendon, every ligament, every part of the digestive system, every part of the lungs, every part of the skin, and every other organ. The wiring has to reach every touch sensor, every pain sensor, every temperature sensor, and every position sensor.

It is impossible to comprehend the level of wiring detail. But to give some idea of its complexity, consider that a Boeing 747 jet has around 170 miles of wiring. That is a lot of wiring, but it's dwarfed by the wiring in the human body, which has some 90,000 miles of wiring.[8] Laid out end to end, that would stretch more than three times around the earth.

Integrating Millions of Nerve Pathways

The nervous system's 90,000 miles of wiring must be precisely integrated with all the other sub-systems of the body. One of the most common nerve pathways involves the motor nerves that travel from the brain to the many thousands of motor units in the body. As noted previously, the total number of motor units in a human body may exceed 100,000. That's a lot of roadway and a lot of crisscrossing. I love looking at anatomical diagrams of a large muscle like the biceps, which show hundreds of muscle motor units with each individual motor unit having its own nerve pathway. Such diagrams allow you to see how each nerve pathway must weave around other motor units to get to its unique motor unit. This level of precision integration is astounding.

As discussed in Chapter 10, among the most common skin sensors are nerve endings that detect movement of body hair. There are several million body hairs on an adult body, and most have nerve endings near the base of the follicle. It is impossible to imagine the intricate details of such wiring. Incredibly, when a tiny insect lands on just one hair on the arm, the brain senses that something has touched the hair. From this the brain senses both that the body has been touched and where.

One thing I find particularly impressive about the wiring in the spinal cord is how the motor nerves are separated from sensory nerves within the spinal cord but then get mixed together after exiting the spinal cord. The reason for separating nerve types in the spinal cord is that motor pathways go to a particular part of the brain called the motor cortex whereas sensory signals all go to another part of the brain called the sensory cortex. The reason for mixing nerve types after exiting the spinal cord is that it reduces the number of bundles going through the body. That is exactly the kind of carefully planned design decision an engineer would make in order to create an efficient system.

A good example of integration is the way bones have holes for nerves to pass through. This reminds me of how I had to put holes in spacecraft structures to get wiring from one part of a spacecraft to another. The skull has more than twenty openings (foramina) to allow nerves and blood vessels to enter and exit the brain. Figure 15.2 shows several of the major openings. The best known of these is the large hole for the spinal cord, called the foramen magnum. But there are other important openings for major parts such as jugular veins and optic nerves. Each of these holes needs to be planned. A hole for the optic nerve from the eye is only useful if there is a functioning eye. But an eye is no use if there is no passageway to the brain for the optic nerve.

A remarkable similarity exists between nerve bundles and communication cables. (See Figure 15.3.) The difference is that nerve bundles are far more sophisticated due to features like an integrated blood supply. Nerve bundles show all the hallmarks of intentional design and planning.

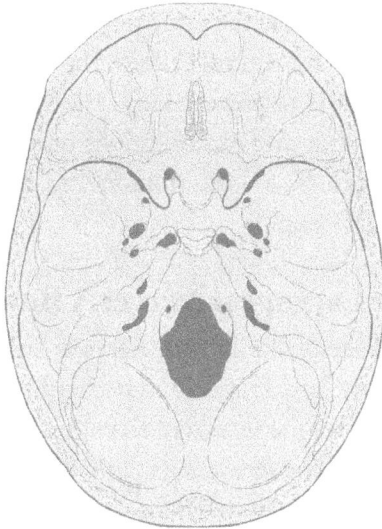

Figure 15.2. Holes in the skull for nerve bundles and veins.

Nerve bundle   **Engineering cable**

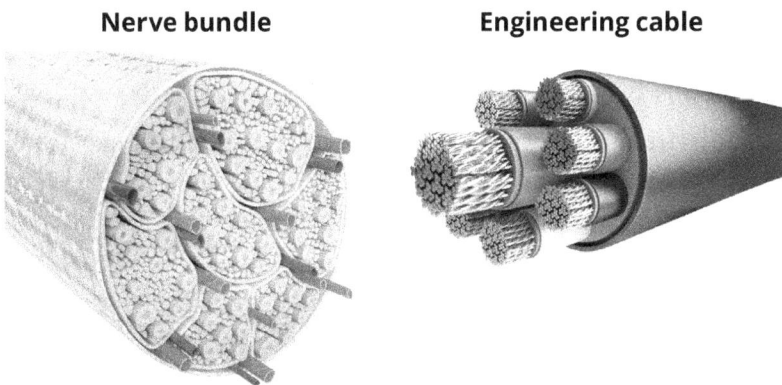

Figure 15.3. Communication cables.

Biomechanics of Control

One reason humans are so much more agile than robots is the body's advanced sensing and control. Our bodies contain thousands of proprioceptors in the joints, which provide position and velocity information. This information is passed to the brain and processed with lightning speed in order to adjust muscle movements to produce stable and efficient action.

For example, when running on uneven ground, the ankles twist and turn, which could cause the body to topple over. But because of the advanced sensing and control system, the brain reacts rapidly to correct joint movements to prevent falls. The brain reacts so swiftly that it is possible to run in dark conditions on uneven ground and, with care and a bit of luck, to keep from falling. Such abilities are far beyond current robot designs.

The Recurrent Laryngeal Nerve: A Botched Design?

Despite the brilliance of the human nervous system, evolutionists such as Nathan Lents, Jerry Coyne, and Richard Dawkins have found fault with it. Lents cites the recurrent laryngeal nerve among his list of "human errors."[9] Coyne sees it as a manifestly defective design due to it taking a seemingly random detour to get to the larynx. "One of nature's worst designs is shown by the recurrent laryngeal nerve," he writes, adding that it "is not only poor design, but might even be maladaptive."[10]

Richard Dawkins piles on. "The recurrent laryngeal nerve in any mammal is good evidence against a designer," he says. "And in the giraffe it stretches from good to spectacular! That bizarrely long detour down the giraffe's neck and back up again is exactly the kind of thing we expect from evolution by natural selection, and exactly the kind of thing we do not expect from any kind of intelligent designer."[11]

Again, I fully agree with these Darwinists that evolutionary theory anticipates poor design in the nervous system. The problem for them is that scientific research points to every part of the nervous system being brilliantly designed, including the recurrent laryngeal nerve.

Two nerve bundles go from the brain to the larynx. There is a superior laryngeal nerve (SLN), which takes a direct path to the larynx. Then there is a second nerve, the recurrent laryngeal nerve (RLN), which drops down towards the heart and then loops back up to the larynx. Lents, Coyne, and Dawkins insist there is no point in having the loop. I find their confidence on this score puzzling, because I have designed loops in wiring systems exactly like the RLN. When I first saw the RLN, the loop struck me as perfectly normal because such loops are quite common in engineering systems. There are at least three reasons for them, all of which apply to the RLN.

(1) To allow additional wire connections

In engineering it is common to join two or more distinct wire bundles together so that they form a larger and more robust bundle. In addition to being more robust, it requires fewer separate routes and fewer support brackets.

The tradeoff is that the larger bundle does necessitate it taking detours to reach all destinations—like a bus taking an indirect route from A to Z to pass all the bus stops. This describes the RLN. It isn't just making a connection to the larynx. It also carries nerve pathways to the esophagus and trachea. It has to make a detour to reach all the destinations, but it makes for a more robust bundle. This is a clever design feature because the disadvantage of the detour is less than the advantage of making a more robust bundle.

I employed this same strategy when designing a spacecraft wiring system. I had several sensor wires piggyback on power wires, creating a large robust bundle. One consequence of this decision, well worth the trade-off, was that the sensor bundle had to perform some looping detours.

(2) To aid assembly

Engineers often use wire loops to make assembly easier. I often designed small loops in my spacecraft wiring to help with assembly. The same advantage applies to the RLN. During embryonic development, the organs move apart, and this can stress the nerve pathways. Having loops can relieve this stress. The looped nerve bundles might also support structures during development.

(3) For redundancy

For applications that require high reliability, like aerospace engineering, it is standard practice to build in redundancy by separating wire bundles into two different routes for each subsystem. If there were only one bundle and it broke, the whole subsystem would be lost. By having two separate routes to the same subsystem, the subsystem is protected from complete failure if one wire bundle is lost. In my spacecraft wiring systems, I always had two separate routes of fully redundant wiring, which greatly improved the reliability of the spacecraft.

The same principle of redundancy applies to the RLN. If one of the laryngeal nerve bundles (SLN or RLN) fails, there is still one left to keep the larynx functioning to some degree. There are examples of soldiers losing either the SLN or RLN after being shot, but they retained the ability to speak thanks to the remaining nerve bundle. This major advantage is only possible because of the detour of the RLN. Lents, Coyne, and Dawkins are clearly unaware of the principles of wiring design and the principle of multi-objective optimization. It is also clear they have not consulted engineers.

The Need for Humility

Since the nervous system is far beyond anything produced by human engineers, and far from being fully understood by scientists, great caution is needed before criticizing its design. A knee-jerk criticism of one tiny part of the nervous system in isolation is likely only to reveal the ignorance and prejudice of the critic and result in a false judgment.

Imagine a high school student taking a quick look at one detail of the wiring on a SpaceX rocket and saying, "That loop looks wrong to me. Wasted wiring. Extra weight. Poor design." That student would be failing to appreciate that a large team of highly skilled engineers spent years perfecting the wiring-system design and that, in consequence, there are probably good reasons for the loop. In the same way, it is inadvisable to act as an armchair critic of this or that feature of the nervous system, such as the recurrent laryngeal nerve, without having a deep understanding of the principles of design and while ignoring the possibility that the feature may be the product of an engineer who is further along than his critic.

90,000 Miles of Evidence

Of all the parts of a complex engineering system, it is often the wiring subsystem that most readily reveals design and planning. This is surely true of the human nervous system. The degree of design detail in this system is astounding and reveals abundant evidence not only of intentional design but of absolutely brilliant design.

To be sure, the nervous system can be compromised by aging, injury, and disease, but no complex engineered system known to man is invulnerable to such threats. The wiring system I designed for the Envisat satellite eventually stopped working because the satellite reached the end of its life after certain components degraded, this after working for more than ten years. The fact that Envisat gave out after ten years does not mean it was badly designed. In the same way, when the human nervous system degrades, that does not negate the argument for design. Any informed and unprejudiced assessment of the human nervous system would judge it among the greatest of engineering achievements, with its only near competitors being other biological systems.

WHICH WATCHMAKER?

Was the designer of life "a blind watchmaker," as Richard Dawkins would have us believe, or a mindful and sighted one, an artificer of unmatched skill? In the preceding chapters, we considered a series of biological marvels that defy evolutionary theory and, as I argued, strongly suggest not merely purposeful design but ingenious design. Those chapters are the heart of this book. If you were to go no further and simply reflect with an open mind on the testimony of those marvelous systems, I would count my effort a success, for a sober look at such wonders has the potential to dissolve all manner of pro-evolutionary misinformation spoon-fed to us since childhood.

But modern evolutionary theory has all manner of arguments and patches at its disposal for attempting to neutralize such evidence for intelligent design. So, in the chapters that follow we transition from examining one biological system at a time to a higher-level investigation of the broader arguments for and against evolutionary theory and intelligent design. I also want to share more details of my experience in advocating intelligent design in academia and the reactions that has caused.

16. Biological Systems Cannot Self-Organize

Engineers Know from Experience

One of my favorite quotations is commonly attributed to the great aerospace engineer Theodore von Kármán: "Scientists discover the world that exists; engineers create the world that never was."[1] I love it because it highlights the fact that design requires immense creativity, insight, and planning. It doesn't just happen. That's why I and many other engineers are put off by the attitude that the first living cell, or the intricate tendon assembly, or any of a thousand other sophisticated biological marvels, must have self-organized because there is no adequate alternative explanation one is willing to consider. That's not a scientific way of thinking. And it's certainly not how I would want someone thinking who was working with me on a challenging engineering project. It is mere wishful thinking, the bane of successful engineering.

As we saw in the first part of this book, biological systems are engineering systems. The difference is that their technology is more advanced than any human technology. Engineers are very aware that their technological systems do not self-organize. To appreciate why this is so, it is helpful to consider some of the challenges of engineering design and also lessons from the history of technology.

Lessons From Spacecraft Design

I was one of the design leaders for the world's largest Earth observation satellite, Envisat. My job was designing the deployment system for the solar panels. The satellite contains over 100,000 complex parts, each with around twenty pieces of critical design information relating to geometry, material properties, and production. That amounts to around two million pieces of critical information for the whole spacecraft. If just one piece of information was wrong, the whole mission could fail. I often explain to students that they might be overjoyed with a mark of 99 percent in a math exam, but that would be utter failure for spacecraft design.

As sophisticated as Envisat was, biological systems are vastly more complex than it or any other manmade spacecraft. Biological systems contain vast amounts of critical information, and because of this there should be precious little tolerance for error. What tolerance that does exist in biological systems often can be traced back to built-in redundancies and sophisticated error-corrections systems, which themselves boggle the mind for their sophistication, and which themselves require an explanation for their origin.

What this suggests is that the complex systems we find in biology must have required for their origin even more leaps of creativity and immense planning than is the case with spacecraft. By extension, this also suggests that the idea that biological systems can self-organize is even more fanciful than believing a spacecraft could self-organize. And no, appeals to biological reproduction and natural selection do not make all the difference.

To illustrate why evolutionary step-by-step processes are no use for creating complex systems, let me share a few details from the design of the solar array deployment system on the Envisat satellite. The twin images in Figure 16.1 show the Envisat satellite before launch and then in space with the solar array deployed.

The solar array consisted of fourteen panels to convert sunlight into electrical energy for the spacecraft. These panels had to be folded into a stack during launch to fit in the rocket nosecone and then deployed in orbit. The solar array was tested on the ground to ensure the panels would unfold and deploy smoothly. The panels were quite large. When unfolded, the total area of the panels was about the size of a badminton court.

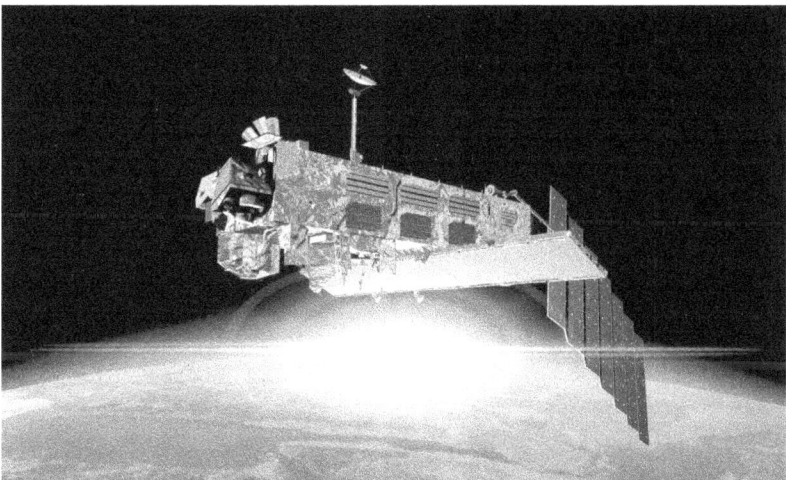

Figure 16.1. Envisat satellite, before launch and in orbit with solar array deployed.

The panel stack also had a robotic arm to move the solar array away from the spacecraft so it wouldn't block the view of the ten instruments on the spacecraft's main body. The solar array was connected to the main body by this robotic arm. The robotic arm was about thirteen feet long and had three hinges functioning as shoulder, elbow, and wrist.

A major challenge for the solar array was designing the gearbox for the robotic arm's three main hinges. As design leader, I knew there was no gearbox in the world that met Envisat's requirements. We needed one that could perform three functions: (1) drive the robotic hinges with a gear ratio like a normal gearbox; (2) withstand movements from launch vibration; and (3) lock the hinges at the end of deployment.

Several of my colleagues from the European Space Agency thought this was an impossible task. That was worrying. As design leader for the solar array deployment system I was under pressure to create a world of gearbox technology that had never existed before. When you have a task of creating new technology, you realize that what's required are gigantic leaps of innovation. Baby steps won't cut it.

After weeks of intense creative thinking, I invented a gearbox called a double-action worm gear, shown in Figure 16.2. It uses the previously existing technology of a worm gear, but it differs from a worm gearbox in there being an additional box which allows the worm shaft to move horizontally as well as in rotation. This double movement is why it is called a double-action worm gearbox. The new gearbox earned a European patent.

The upper image in Figure 16.2 shows the gearbox in the launch and deployment mode. The lower image shows the gearbox in the locked orbit mode. The unique gearbox can perform the motion of three different machine elements. During launch the gearbox behaves like a rack and pinion gear. During deployment the gearbox acts like a worm gear. Then at the end of deployment the gearbox acts like a screw-nut.

The gearbox is a good example of how engineers do not use a Darwinian process to develop new technology. To change a standard worm gearbox into the double-action worm gearbox requires hundreds of simultaneous changes. The double-action worm gearbox cannot be brought into being through single changes, not with functional results each step of the way. Plus, the engineer doesn't settle on all the particular innovations until he envisions the completed new system with all the novel parts and arrangements working in concert. The same holds for biological systems: They cannot evolve tiny step by tiny step, one small mutation at a time. They couldn't if guided by an engineer, and they would be doubly incapable of doing so if the development process were "flying blind" into the future, Darwinian style, unguided by any future design goal and with the punishment for failure at any stage being death.

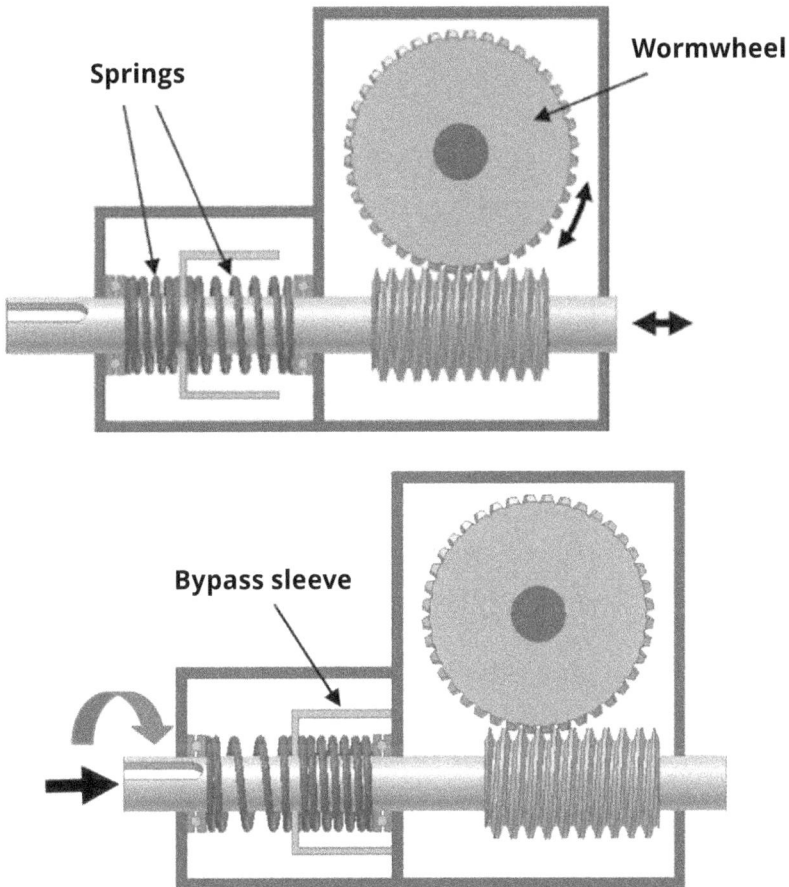

Figure 16.2. Double-action worm gearbox.

The invention of sophisticated engineering marvels requires many skills: an intimate knowledge of physics, an understanding of how to manipulate physical effects with components, a creative mind, and determination. But invention is only half the battle. After invention comes the long task of modeling, testing, and refining.

Figure 16.3 shows one of the smaller efficiency equations I had to work out for the gearbox. Since the gearbox was new, I could not just look the equations up in a book or online. They had to be developed from scratch. And not to complain, but that was hard. I sometimes joked that there were only two people who understood those equations—myself and the reviewer of my paper; and I was never sure if even the reviewer did.

$$T_{\text{OUT}} = \frac{1}{1 + \mu_K} \tag{b}$$

$$\times \left\{ \frac{T_{\text{IN}}\,\eta_{\text{BACK-T}}}{N} \right. \tag{a}$$

$$- \left(P_U - F_Z \frac{K_1}{K_1 + K_2} \right) R_F\,\mu_F \tag{c}$$

$$- \left(P_U + F_Z \frac{K_2}{K_1 + K_2} \right) R_{\text{BA}}\,\mu_{\text{BA}} \tag{d_1}$$

$$- F_Z\,R_{\text{BA}}\,\mu_{\text{BA}} \tag{d_2}$$

$$- P_L\,R_{\text{BA}}\,\mu_{\text{BA}} \tag{d_3}$$

$$- (F_Z - P_L)R_{\text{BR}}\,\mu_{\text{BR}} \tag{e}$$

$$- (F_Y^2 + F_X^2)^{1/2} R_J\,\mu_J \tag{f}$$

$$- (F_Y^2 + F_X^2)^{1/2} \frac{R_{\text{WH}}\,\mu_J\,\eta_{\text{BACK-T}}}{N} \tag{g}$$

$$- (F_Y^2 + F_Z^2)^{1/2} \frac{R_J\,\mu_J\,\eta_{\text{BACK-T}}}{N} \tag{h}$$

$$\left. - \frac{F_X\,R_F\,\mu_F\,\eta_{\text{BACK-T}}}{N} \right\} \tag{i) \quad (19a}$$

Figure 16.3. An efficiency equation for my double-action worm gearbox invention.[2]

The requirements for the gearbox were extremely challenging. In space the gearbox would face sub-freezing temperatures and scorching heat, depending on whether the robotic arm was in shadow or sunlight. The vacuum of space was a major challenge because it ruled out use of wet lubrication. Another challenge was there were three hundred solar array power lines across each hinge. This meant the actuators had to overcome the resistance from these wires. Another challenge was the zero-gravity environment of space. Zero gravity was impossible to replicate on the ground, so we had to rely on what we hoped was very accurate modeling to check the deployment dynamics.

Before building the prototypes of the new double-action worm gearbox, I presented the concept to the engineering department of Oxford University. I was surprised (and a little concerned!) when the leading engineering design professor insisted it would not work. He could not believe that one gearbox could perform three kinematic functions.

The Nerve-Wracking Moment of Truth

Several years before launch, the Envisat project manager called me in for a meeting. He asked if I realized that the mission cost £2 billion. I replied that I had not known that. He then asked if I realized that if "your solar array" did not deploy within one hour, the whole mission would fail. I replied again that I had not known that. He then asked me to remember that for the rest of the project! At that point I knew that my gearbox simply had to work, and work the first time. While I could build in some redundancy, such as having two drive motors and two sets of wiring, there were some parts, such as the gears, where this was impossible.

In the wee hours of the morning on March 1, 2002, I watched the launch of the £2 billion satellite, sent into orbit via an Ariane 5 rocket. Waiting to see if the solar arrays would deploy was the longest and most nerve-wracking ten minutes of my life.

Many people have asked how I felt before and during launch. My answer is that, like most spacecraft designers, I felt physically sick and had not slept properly for several nights. I knew there were tens of thousands of design details in the solar array and that everything had to work for a successful deployment.

Thankfully, on that early March morning I had tears of joy because the solar array worked perfectly and Envisat became the largest Earth observation satellite in the world. Over the next ten years Envisat would produce some of the most spectacular pictures of Earth ever taken. Following the success of Envisat, my gearbox was used on the European METOP series of satellites.

For my work on Envisat I was awarded several national prizes, including the Turners Gold Medal and the James Clayton Prize, the top mechanical engineering award in the UK. I am hoping to bump into the Oxford professor one day to tell him my gearbox did work!

I toot my horn here to help drive home a point: As successful as I've been blessed to be as an engineer, when I look at biology, I see systems far more intricate and advanced than anything I or other engineers have ever designed. In the world of life I find systems involving extraordinary leaps of complex technology, ones that required immense creativity and planning, systems that could not evolve step by step or self-organize.

Harnessing Physics Involves Immense Innovation

A key reason engineering design is so difficult is that you have to harness networks of complex physical effects (e.g., spring, lever, combustion) in order to fulfill complex functions like producing power and torque from an engine. To harness physical effects it is first necessary to conceive of complex assemblies of interacting components that will produce the desired combinations of physical effects.

Take, as an example, the complex engineering system known as the internal combustion engine. This kind of engine is so familiar to us today that it might be tempting to think it was not especially difficult to design. In fact it was extremely challenging to do so. The designers started with the goal of converting fuel combustion into mechanical rotary motion. Through ingenious thought, designers realized that controlled combustion in a closed cavity would increase pressure and push a piston along a cylinder. The designers also realized that if the piston was connected to a crankshaft, then the linear motion would be converted to rotary motion.

Another key innovation was to employ valves operated by the crankshaft to introduce the fuel into the cylinder, and to draw the exhaust out at the optimal time. The designers also realized that if several pistons were connected to the same crankshaft at different rotary positions, then the shaft would be powered in a smooth way.

There are various other elements of technology used in internal combustion engines that contribute to efficiency, safety, and control. Yes, these engines have "evolved" over the years, but each new innovation required intelligence, foresight, planning, and a host of adjustments to the whole. A typical modern internal combustion engine for a car contains over 200 precision interacting parts, and thousands of pieces of information are used for manufacture. The internal combustion engine is a triumph of creative engineering. It is also a classic example of irreducible complexity, in which many parts are needed to create a functioning system.

The really difficult part of design is having the vision to see the interconnectivities, such as using the same crankshaft to perform several different sub-functions. In addition to harnessing many physical effects to produce a desired function, the designers must deal with unwanted physical effects that cannot be avoided, such as excessive thermal expansion or unwanted wear and noise.

Part of what makes design so challenging is that the design space is practically infinite, meaning there are a countless number of possible configurations. The vastness of the space is due in part to the hundreds of physical effects engineers can employ. There are physical effects in areas like solid mechanics, aerodynamics, fluid mechanics, thermodynamics, electrical systems, electronics, optics, and acoustics. There is also an endless number of ways to use materials and component geometries to produce these physical effects. The resulting design space, oceanic in size, means the designer has to be decisive, clever, and fast in navigating the space and homing in on a solution.

Achieving adequate levels of fine-tuning is another major design challenge. The piston of an internal combustion engine must have an extremely precise fit inside the engine cylinder to function. Other details that must be precise include the valve timings, the combustion timing, the cooling system, and the air-fuel ratio.

A complex system such as this could not evolve one small functional step at a time. Yes, the internal combustion engine has "evolved," but not through any mindless or gradual process, and the innovations were themselves leaps, requiring the simultaneous addition of multiple new parts and the careful reconfiguration of existing parts. The same applies to biology. It can be tempting to think that biological systems evolved step by tiny step, pushed along by random mutations and guided by natural selection. But that ignores the hard reality that great leaps of creativity are needed to achieve novel function, with a vast suite of new and innovative details required to come online simultaneously.

History Demonstrates the Immense Challenge of Innovation

The internal combustion engine is not unusual in requiring huge leaps of creativity for its origin. Virtually every technological innovation since the birth of science, and many before that, have involved irreducible complexity in spades.

The following list contains sixteen key inventions and their approximate dates:

- Mechanical clock, circa 1300
- Piston, 1680
- Piston-and-cylinder steam engine, 1712
- Airfoil, 1804
- Electric motor, 1821
- Mass-production of steel, 1856
- Internal combustion engine, 1872
- Telephone, 1876[3]
- Safety bicycle, 1885

- Controlled aircraft, 1903
- Affordable motor car, 1908
- Jet engine, 1930
- Programmable computer, 1941
- Silicon transistors, 1954
- Internet, 1983
- Smartphone, 1994

Notice that it took a hundred years to get from an airfoil to an airplane. It took more than a hundred years to get from the first telephone to the first smartphone, two hundred years to get from steam engines to widely affordable motor cars, and seven hundred years to get from mechanical clocks to smartphones. These long periods of time reflect the immense challenge of design. The slowness doesn't stem from a lack of motivation. The importance of technology in both commerce and warfare means designers have been extremely motivated and well paid. Or to take a specific example, people have recognized time out of mind the benefits of producing an engine that could efficiently convert fuel into mechanical work. Yet it was not until 1712 that Thomas Newcomen made the spectacular breakthrough with the invention of the piston-and-cylinder steam engine.

The reason for the slowness has all to do with the great difficulty of design. So many aspects have to be considered, including materials, manufacture methods, analysis methods, and assembly methods. Each area requires considerable creative progress, with countless ways to go wrong.

Selling Evolution with Pictures

Books on evolution often include picture sequences that make the idea of blind evolution inventing complex systems seem plausible. Richard Dawkins's *The Blind Watchmaker* provides an example. The book features a two-dimensional picture gradually morphing from one type of shape into another. Dawkins then argues that if there was a selective advantage for the latter type of shape, it would eventually get selected.

However, Dawkins's illustration ignores the fact that engineering systems are not pictures on a page but assemblies of hundreds of precisely shaped three-dimensional components with complex interactions. While pictures can change incrementally through tiny changes, complex machines require hundreds of simultaneous design changes to be transformed into a different kind of complex functional system.

The evolutionist's belief requires him to ignore facts obvious to anyone with even basic knowledge of engineering and biology. These facts include:

- Technological worlds that have never been are created not by blind forces but by engineers exercising foresight, creativity, and planning.

- Novel complex systems require huge leaps in creativity, often involving decades or centuries of striving, along with the input of multiple geniuses.

- The resulting complex systems require exquisite fine-tuning.

- We never witness novel complex systems originating through self-organization. They require an enormous amount of intelligent planning.

- It takes many years and considerable talent to qualify as a systems designer due to the complexity and demands of designing complex systems.

- The complex systems found in our most advanced human technology pale beside the complex systems we find in biological systems.

Evolution may seem plausible through picture animations, but pictures are wholly inadequate to represent complex three-dimensional interacting parts. How intelligent people such as Richard Dawkins could mistake relatively simple pictures for the profoundly complex biological reality that the pictures are meant to represent is a question for historians of science and students of human nature. But one can hazard a guess that the oversight bespeaks the power of a reigning scientific paradigm to mold thought and, beyond that, the power of the materialistic worldview to render the live alternative unacceptable.

The Difficulty of Reaching System-Level Designer

One way to appreciate the challenge of design is to consider how difficult it is to become a designer of complex systems. It can take decades of training and experience to become competent at designing complex systems like those in spacecraft. A typical route to becoming a systems designer starts with a four- or five-year master's degree in engineering. That degree includes studying math and physics to a relatively deep level as well as other subjects like computer science, computer-aided design, and materials science.

Despite studying engineering for four or five years, a graduate is not yet ready to design complex systems. It is necessary to gain experience in design at the component level and then the subsystem level before moving on to system-level design. In fact, most engineers never reach system-level design. Only the most talented engineers get to perform system-level design, and it often takes them years and years to reach that level, because, again, designing novel complex systems is mind-bendingly difficult. So, then, why are we to believe that mindless evolution has managed the trick countless times?

I can see how someone who has not designed complex systems might be tempted to think complex systems can self-organize. I can also see how someone who thinks "evolution must have happened because we are here" will believe that complex systems must be able to self-organize. But I would say to such people, try and design a complex system for yourself and then you will be much less inclined to believe in self-organization.

17. Evolutionary Theory in Decline

Evolution: Proven Fact or Embattled Theory?

In his book *The Greatest Show on Earth*, Richard Dawkins insists evolution is a proven fact. Students often hear the same thing in biology classes, and are told that denying it is roughly on par with denying a round Earth. But when I discuss evolution with my academic colleagues, they readily admit it is a controversial theory.

Of course, no one questions the two basic tenets of (1) change over time in the history of life and (2) microevolution—e.g., changing beak sizes in finches. However, when it comes to the origin of life or macroevolution or human origins or sexual selection, there is actually considerable controversy within academia.

When you look in college textbooks to see the "hard proof" of evolution in action, you find it is primarily evidence for microevolution. If one continues to dig into the mainstream, peer-reviewed scientific literature, the next thing that becomes clear is that key aspects of evolutionary theory's purely naturalistic story—extending from the origin of the first life through the rise of *Homo sapiens*—are often treated as controversial in the mainstream literature, and that among the naysayers are top mainstream scientists. True, any one of these naysayers focuses on only one or two problems with the theory, but collectively these establishment naysayers call into doubt most or all of the major lines of evidence for the theory.

234 / Ultimate Engineering

This chapter will explore this controversy that students are routinely assured does not exist, and will cover both the origin of life and biological evolution. Some might object that Darwin's theory did not consider the origin of life. But Darwin theorized in private about a purely naturalistic explanation for the first life, and the larger project of modern evolutionary theory includes both biological and chemical evolution, both the origin of life's diversity of forms and the origin of the first life. So we will examine both, taking them together as complementary parts of that full-orbed project that is modern evolutionary theory.

Public Confidence, Private Doubts

The public-facing scientists who champion evolutionary theory typically project a supremely confident face to the world, bristling at the charge that it's a mere theory that a reasonable person might question. But some scientists paint a different picture.

The distinguished scientist Paul Davies, for example, states frankly, "Many investigators feel uneasy about stating in public that the origin of life is a mystery, even though behind closed doors they freely admit that they are baffled."[1] His comment doesn't surprise me, because over the past thirty years I have regularly asked biologists in private what they actually believe about abiogenesis. Repeatedly they have confessed that, although they would never want to admit this to the public or their students, there is precious little evidence for the theory. As one of them, a non-religious senior professor of microbiology, memorably put it to me, the field is no better than "black magic."

As for the theory of biological evolution, which takes matters up after the origin of the first life and is often referred to simply as Darwinism or evolution—it's rare to get a card-carrying proponent to admit that the paradigm is, as geneticist Michael Denton put it, "a theory in crisis." But if you assemble all the bits of frank public talk from the lot of them, a much less confident picture emerges.

Take the fossil record. It is well known among specialists that it does not support the Darwinian story of gradualistic evolution of all life from a common ancestor. As the eminent Harvard paleontologist

Stephen Jay Gould put it, "The extreme rarity of transitional forms in the fossil record persists as the trade secret of paleontology."[2] He made that comment half a century ago, but it remains true today. (See the discussion below under the subheading, "Contrary Evidence from the Fossil Record.")

What about Darwin's idea of a gradually branching tree of life, regularly presented as settled fact in biology textbooks? Graham Lawton reported at *The New Scientist*:

> "For a long time the holy grail was to build a tree of life," says Eric Bapteste, an evolutionary biologist at the Pierre and Marie Curie University in Paris, France. A few years ago it looked as though the grail was within reach. But today the project lies in tatters, torn to pieces by an onslaught of negative evidence. Many biologists now argue that the tree concept is obsolete and needs to be discarded. "We have no evidence at all that the tree of life is a reality," says Bapteste. That bombshell has even persuaded some that our fundamental view of biology needs to change.[3]

And here is German geneticist Günter Theißen, concerning the dearth of evidence for macroevolution: "While we already have a quite good understanding of how organisms adapt to the environment, much less is known about the mechanisms behind the origin of evolutionary novelties, a process that is arguably different from adaptation. Despite Darwin's undeniable merits, explaining how the enormous complexity and diversity of living beings on our planet originated remains one of the greatest challenges of biology."[4]

The influential American evolutionary biologist Lynn Margulis and her co-author, Dorion Sagan, said something similar, but at greater length and even more emphatically:

> We agree that very few potential offspring ever survive to reproduce and that populations do change through time, and that therefore natural selection is of critical importance to the evolutionary process. But this Darwinian claim to explain all of evolution is a popular half-truth whose lack of explicative power is compensated for only by the religious ferocity of its rhetoric. Although random

mutations influenced the course of evolution, their influence was mainly by loss, alteration, and refinement. One mutation confers resistance to malaria but also makes happy blood cells into the deficient oxygen carriers of sickle cell anemics. Another converts a gorgeous newborn into a cystic fibrosis patient or a victim of early onset diabetes. One mutation causes a flighty red-eyed fruit fly to fail to take wing. Never, however, did that one mutation make a wing, a fruit, a woody stem, or a claw appear. Mutations, in summary, tend to induce sickness, death, or deficiencies. No evidence in the vast literature of heredity changes shows unambiguous evidence that random mutation itself, even with geographical isolation of populations, leads to speciation.[5]

And no, these scientists are not giving up on evolutionary theory, much less endorsing intelligent design; but that's precisely the point. These are mainstream evolutionists, committed to finding a fully materialistic account for the origin and diversification of all life; *and yet* they are admitting such things.

The No-Intelligence-Allowed Mandate

The problems highlighted in these quotations are only a small sampling of the deep evidential problems confronting the paradigm. So, why don't these scientists just give up on evolutionary theory? Of course each scientist is unique and is likely guided by a cluster of motivations unique to him or her. But from thirty years of talking with such scientists privately, my sense is that one strong motivation for many of them is this: They define science as necessarily materialistic (without sound justification) and are then inevitably convinced that if our biosphere did in fact arise through purely material forces, then some version of gradualistic evolution must be true, since no other materialistic scenario is remotely as plausible as evolutionary gradualism.

It's hard to overemphasize the role of materialistic philosophy in their thinking. Harvard geneticist Richard Lewontin frankly admitted that he and many of his fellow scientists are dogmatically committed to materialism, to the point of equating the materialistic framework with science itself, and of embracing evolutionary "just-so stories" even when those stories lack confirming evidence. He summarized this outlook in the *New York Review of Books*:

> We take the side of science *in spite* of the patent absurdity of some of its constructs,… *in spite* of the tolerance of the scientific community for unsubstantiated just-so stories, because we have a prior commitment, a commitment to materialism. It is not that the methods and institutions of science somehow compel us to accept a material explanation of the phenomenal world, but, on the contrary, that we are forced by our *a priori* adherence to material causes to create an apparatus of investigation and a set of concepts that produce material explanations, no matter how counter-intuitive, no matter how mystifying to the uninitiated. Moreover, that materialism is absolute, for we cannot allow a Divine Foot in the door.[6]

The influential evolutionist Sir Julian Huxley expressed the same unbending rule in a single sentence: "Modern science must rule out special creation or divine guidance."[7]

Thus does the evolutionary paradigm graduate to dogma, with its adherents consciously and conscientiously refusing to follow any evidence that contradicts evolutionary materialism.

The above overview is intended merely to show that whereas evolutionary theory is presented to the public as settled fact, several leading evolutionists have conceded in their franker moments that key parts of the theory are starved for evidence, and that the theory is being propped up by a dogmatic commitment to materialism. It would seem that many evolutionists promote evolutionary theory as fact not because of the evidence, but despite it. Let's slow down now and look more closely at why both the origin of life and macroevolution face serious trouble from the evidence.

Declining Evidence for Abiogenesis

Abiogenesis refers to the idea of the first life emerging purely naturalistically—no designer needed. For abiogenesis to occur, lifeless chemicals would have to spontaneously assemble into organic components that would, in turn, have to assemble themselves into incredibly complex structures, which would have to combine with many other complex interacting parts to produce complex machinery such as information systems and energy systems. And all these biological machines would have to function as part of an integrated system that could reproduce, repair, and fuel itself—something no manmade machine is anywhere close to being able to achieve.

At one time people believed that life routinely arose from non-life—spontaneous generation. But scientific experiments, culminating in one by Louis Pasteur, put an end to such notions. Shortly after Darwin's *Origin of Species* appeared in print, it was settled that at least in the ordinary course of things, life only comes from life. That left Darwin to conjecture in a private letter to his friend Joseph Hooker that perhaps, given just the right extraordinary circumstances on the early Earth, in some "warm little pond," life could have spontaneously arisen from non-life.[8] Alfred Russel Wallace, credited with proposing the theory of evolution by natural selection at the same time as Darwin, wasn't convinced. He thought the origin of life, along with the origin of humanity, were each a bridge too far for blind evolution. He made the argument in a book titled *Darwinism*, and he later summarized what the book urged regarding the limits of the theory of evolution by natural selection: "My argument in *Darwinism* was to show that there were peculiarities in our mental nature that could not be explained by a development through the law of 'natural selection.'"[9]

Supporters of Darwin and scientific materialism hoped that one day Wallace would be proven wrong. But modern discoveries have only increased the evidence in Wallace's favor.

The great scientist Lord Kelvin also made it clear that he thought abiogenesis was impossible. He said, "Was there anything so absurd as to believe that a number of atoms by falling together of their own accord could make a sprig of moss, a microbe, a living animal?"[10] Kelvin made this statement long before scientists knew about the breathtaking complexities of the cell, a knowledge that only makes matters worse for the theory of abiogenesis.

Lord Kelvin's deep knowledge of thermodynamics is relevant here. The second law of thermodynamics involves a principle that systems tend toward disorder. The problem with the theory of abiogenesis is that it proposes that great disorder moved spontaneously to immense order.[11] So it is no wonder Kelvin was skeptical.

The Miller-Urey experiment of 1953 is often cited in school textbooks as evidence to support abiogenesis. But all Stanley Miller and Harold Urey made were a few amino acids, and even then it was under conditions that are no longer thought to accurately reflect the early Earth. Among origin-of-life scientists it is well known that the experiment does not represent more than, at best, an inch of progress down a football field where the goal posts have been receding farther and farther into the distance with every passing decade as more and more layers of cellular complexity are discovered. This lack of progress is despite many thousands of attempts and using some of the best labs in the world. Scientists haven't even succeeded in producing life in the lab de novo, even with millions of dollars spent and even allowing themselves all sorts of intelligent interventions in the process to counteract nature's relentless tendency toward entropy.

The year 1953 is significant for another reason. It is the year James Watson and Francis Crick discovered the helical structure of DNA. In the decades following this revelation, more and more of the intricacies of the cell were discovered, including the genetic code and various shockingly clever molecular machines such as kinesin proteins, which walk along structures in the cell.

"An honest man," commented Crick in 1981, "armed with all the knowledge available to us now, could only state that in some sense, the origin of life appears at the moment to be almost a miracle, so many are the conditions which would have had to have been satisfied to get it going."[12] In the more than forty years since Crick made that remark, our understanding of just how astonishingly complex "simple" cells really are has only grown, and exponentially so.

Crick, notice, hedges with the phrase "almost a miracle." He was committed to scientific materialism, so he wouldn't allow himself to let Lewontin's forbidden "Divine Foot in the door." But even Crick had to admit that there was no naturalistic explanation and that it sure looked as if a miracle was required for the origin of the first life.

Paul Davies summarized the problem: "Just as bricks alone don't make a house, so it takes more than a random collection of amino acids to make life. Like house bricks, the building blocks of life have to be assembled in a very specific and exceedingly elaborate way before they have the desired function."[13]

And that is putting it mildly, since the set of "bricks" involved in building a living cell are vastly more complex than ordinary bricks. A large, dynamic engineering system such as a car would be a somewhat less simplistic illustration. A car contains thousands of components precisely assembled. If someone found a way of producing lumps of aluminum and steel, this would not represent any meaningful progress towards creating a functioning motor car by random processes, all the more so given that even the tiny step of producing the aluminum and steel required intelligent intervention.

Declining Evidence for "Simple" Organisms

At the risk of seeming to pile on, be aware that even relatively simple cells dwarf an automobile in complexity. And unlike an automobile, each such cell also functions as a factory that can build copies of itself—a factory of factory-building factories.

If life began with very simple organisms containing small amounts of genetic information, then we would expect to find evidence of such organisms, either still in existence or preserved as fossils. Scientists have dated fossilized microorganisms in Western Australia to 3.5 billion years ago, and in Canada to at least 3.77 billion years ago. But these and other ancient microbial fossils stubbornly refuse to provide evidence of truly simple microbes. Ideally, researchers would find an organism, living or fossilized, with a small genome of perhaps fewer than a hundred units of information. But the scientific evidence, on multiple levels, tells us that there are no such simple organisms.

To monitor progress in the search for simple life, researchers record new discoveries of the smallest known genome. In 2022 there was a report about the discovery of the smallest known genome on the LiveScience website: "A bacterium living in special cells inside an insect has the smallest genome of any known cellular lifeform, a new study finds.... The genome of *Carsonella ruddi* is less than half the size thought to be the minimum necessary for life."[14]

Despite breaking the record for the simplest lifeform, the bacterium contains a voluminous 160,000 base pairs of information, vastly more than one hundred. Think about it: 160,000 base pairs is comparable to the information required to build a car. Also keep in mind, this is a parasitic organism. The biological information required for the simplest freestanding, non-parasitic microbe is probably greater. But even setting that likelihood aside, the idea that an organism with 160,000 base pairs of information could spontaneously appear by some combination of chance and law-like processes is like believing that a brilliant software program of similar length could come into being without the designing mind of a programmer. It is not just improbable; it is impossible.[15]

Another key test for the theory of abiogenesis is to take the simplest known organisms and modify their genomes to create something simpler. The idea is to delete everything from the DNA that is not essential. The resulting organism is called the minimal cell. Given that scientists can choose any kind of bacterium or similar organism and prune its genome in the lab, it should be possible to create a simple functional cell if such an entity were possible. But as it turns out, scientists have not been able to create a cell that is simple.

In 2016 the first man-made genome of a minimal cell was created from the natural parasitic bacterium *Mycoplasma mycoides*. It was called bacterial cell JCVI-Syn3.0. The modified organism performs only the essential cellular functions common to all organisms, such as DNA replication, RNA transcription, protein translation, and cell division. Even though JCVI-Syn3.0 performs just the simplest functions, it has 531,490 base pairs in the DNA, 438 protein-coding genes, and 35 RNA-coding genes.[16] This finding, that scientists cannot strip down an organism to create a simple functional organism, adds more weight to the conclusion that abiogenesis is impossible.

A third line of evidence comes from engineering efforts to build a self-replicating machine—that is, a machine that can create machines like itself that can create machines like themselves, ad infinitum, in the way that a living cell can. Engineers are nowhere close to achieving this goal. But the research efforts to that end have driven home just how astonishingly complex a self-reproducing machine has to be in order to work.[17] This is relevant to origin-of-life studies because for Darwinian natural selection to kick in, you first need a self-reproducing entity. So nature has to give us an unfathomably sophisticated entity, one capable of self-reproduction, before it can even avail itself of Darwinian natural selection.

Questionable Confidence

In the face of this mounting evidence against a purely naturalistic origin of the first life, the abiogenesis research community, most of whom are committed to evolutionary materialism, go right on whistling past the graveyard.

One example of misleading confidence appears in the UK national curriculum. University of Manchester physicist Brian Cox tells students that hydrothermal vents have "everything inside these vents" needed for the origin of life, because they contain energy and molecules found in organic matter (biogenic materials).[18] He explains that energy and biogenic materials are released when acidic seawaters mix with the alkaline water flowing from a hydrothermal vent. The clear impression given is that students can be fully confident that such a process has everything required to create life.

What Cox does not mention is that the presence of energy and biogenic materials is monstrously insufficient to produce the first life, since what remains missing is any way to assemble the materials into a self-reproducing system—a system that, as we have seen, is necessarily sophisticated in the extreme. Mixing organic matter, acidic seawater, and alkaline water is no more sufficient for the task than is a pile of raw materials (e.g., metal, plastic, synthetic rubber), a lightning bolt, rainwater, and a tidal wave for building a car.

Yet another example of unwarranted confidence in abiogenesis is found in the field of astrobiology. In recent years there have been claims that perhaps asteroids delivered biogenic materials that seeded life on Earth. For example, a 2025 astronomy journal paper reported finding abundant ammonia and other building blocks of life on an asteroid,[19] and reports of this research were quick to state that this could explain the origin of life on Earth. A popular news report that same year outrageously claimed, "Human DNA detected in 2 billion year old meteorite,"[20] even though no such discovery had occurred; all they found was carbon, ammonia, salts, and amino acids in an asteroid. At least the news article acknowledged that the reason for seeking life in space is that "we don't know why" life exists on Earth. But in both stories, there was no mention of the fact that there is no evidence that these raw materials can assemble themselves into a living system, and abundant evidence that they cannot.

Microevolution Does Not Cause Macroevolution

The myth has it that when Charles Darwin visited the Galápagos islands on his globe-circling Beagle voyage as a young man, he observed that the beaks of the finches had changed due to natural selection, eventually triggering a eureka moment. The reality is that Galápagos finches made little impression on him and played little if any role in his formulating the idea of evolution by natural selection. Later Darwinists seized on the observation that the size of Galápagos finch beaks vary over generations in response to droughts and wetter periods, and then read it back into Darwin, his Beagle voyage, and his theory of evolution.[21]

Finches thus became an icon of evolutionary theory, with this evidence of anatomical plasticity in the finch held up as powerful evidence that these and other changes to the bird could accumulate over numerous generations and change the finch into another kind of creature or, conversely, had changed a distinct ancestral creature into a finch. In sum, Darwinists hold up the Galápagos finch as a snapshot of macroevolution in action.

However, it is now well known that finch beak variation occurs only within a narrow range and that the Galápagos finch populations of varying beak sizes are better viewed as subspecies, with all the finches perfectly capable of interbreeding.[22] Thus there is nothing here approaching a process that could accumulate a plausible series of genetic mutations to change the finch into another type of creature, or evolve a fundamentally different creature into a finch.

Following the discovery of genetics in the twentieth century, it became all the more apparent that there is a vast difference between microevolution (superficial changes through adaptation) and macroevolution (major changes involving the origin of fundamentally new organs, systems, and body plans). Whereas microevolution involves little if any creation of new information,[23] macroevolution would require the creation of massive amounts of new, functional information.

Microevolution, such as changes in the color of a moth, is a well-established fact of science. No sane and modestly informed naturalist before or after Darwin has doubted it. However, the idea of macroevolution, such as the claimed evolution of whales from land mammals, is speculative, has very little scientific evidence for its support, and much scientific evidence arrayed against it.[24]

Understand, doubts about microevolution's ability to accumulate into macroevolutionary change are not restricted to the minds of wild-eyed creationists. "A long-standing issue in evolutionary biology is whether the processes observable in extant populations and species (microevolution) are sufficient to account for the larger-scale changes evident over longer periods of life's history (macroevolution)," commented University of Maryland developmental evolutionary biologist Sean Carroll in the journal *Nature*. "Outsiders to this rich literature may be surprised that there is no consensus on this issue, and that strong viewpoints are held at both ends of the spectrum, with many undecided."[25]

Finding quotations from mainstream biologists that emphasize the limits of microevolutionary adaptation is not difficult. Here, for example, is evolutionary developmental researcher Scott Gilbert being quoted in the journal *Nature*: "The modern synthesis is remarkably good at modelling the survival of the fittest, but not good at modelling the arrival of the fittest." The same article quotes paleobiologist Graham Budd as saying, "When the public thinks about evolution, they think about the origin of wings and the invasion of the land. But these are things that evolutionary theory has told us little about."[26]

Or consider this remark from leading paleontologists Douglas Erwin and James Valentine in the *Proceedings of the National Academy of Sciences*: "The appearance of many novel morphologies, frequently expressed taxonomically as new phyla, classes, or orders, occurs with such rapidity in evolutionary time that microevolutionary substitutions involving structural genes seem an implausible mechanism."[27] The following year they doubled down: "We conclude that the extrapolation of microevolutionary rates to explain the origin of new body plans is possible, but does not accord with the primary evidence."[28]

A trio of scientists in the journal *Developmental Biology* were even more plainspoken. "Genetics might be adequate for explaining microevolution, but microevolutionary changes in gene frequency were not seen as able to turn a reptile into a mammal or to convert a fish into an amphibian," they commented. "Microevolution looks at adaptations that concern only the survival of the fittest, not the arrival of the fittest. As Goodwin (1995) points out, 'the origin of species—Darwin's problem—remains unsolved.'"[29]

The problem has grown so acute that a growing chorus of voices from within mainstream evolutionary biology are calling for a radical revision of evolutionary theory. Paleontologist Günter Bechly, a globally recognized expert in fossil insects and former curator at the State Museum of Natural History in Stuttgart, Germany, reported on these efforts. His analysis is worth quoting at length:

> The crucial problems of neo-Darwinism are meanwhile well known and acknowledged in mainstream science, which is why there is a growing trend among theoretical biologists towards a so-called extended evolutionary synthesis. This was also called a Third Way of Evolution, alluding to two other ways: Neo-Darwinism which is considered as a failure, and intelligent design theory, which is a route mainstream academia does not want to go. One of the main proponents of an extended synthesis is the Austrian scientist Prof. Gerd Müller, who held a very revealing keynote at the conference *New Trends in Evolutionary Biology* at the prestigious Royal Society in London in November 2016. In his keynote Professor Müller showed a slide that introduced five explanatory deficits of the

Modern Synthesis, which is just a synonym for Neo-Darwinism (Müller 2016, Bechly et al. 2019). Among these five explanatory deficits, which Neo-Darwinism fails to explain, he listed phenotypic complexity, phenotypic novelty, and non-gradual forms of transition. If Neo-Darwinism cannot explain these crucial phenomena, then it miserably fails as sufficient explanation for macroevolution in general. Alternative mechanisms suggested by the proponents of the extended evolutionary synthesis (e.g., niche construction, phenotypic plasticity, hybridogenesis, natural genetic engineering, evo-devo, epigenetics, GRNs, etc.) cannot overcome the explanatory deficits either, because they either do not address the crucial problem of the origin of novel complex specified information (CSI), or require Neo-Darwinian mechanisms to explain their own origin (e.g., evolvability). If Neo-Darwinism fails on mathematical and empirical grounds, then these alternative mechanisms could never have originated and are dead in the water.[30]

Unfortunately, many high school and college textbooks make no mention of these grave difficulties facing evolutionary theory and instead give the misleading impression that it's a settled fact that microevolution has virtually unlimited metamorphic powers.

The Body-Plan Hurdle

To understand the limitations of adaptation, it is helpful to recognize that the design of an organism involves thousands of design parameters. Examples include leg length, leg joint positions, and leg bone positions. Whereas adaptation can lead to a change in the value of a parameter (e.g., lengthen a leg), adaptation cannot create new parameters and new systems. Adaptation can change the shape and size of a body plan; it cannot change the body plan. To give a hypothetical example, adaptation could not change a marsupial-like mammalian form into a placental mammal, or vice versa, because the two creatures have fundamentally different design parameters. Marsupial mammals have unique parameters that define the pouch design, and placental mammals have unique parameters that define the placental design.

The way engineers design systems like cars illustrates why micro-evolution cannot change body plans. Once a car type—say, an SUV—is designed, engineers list all the parameters and design rules that define the value of those parameters. If there is then a desire for a variant on the SUV design, such as a larger version, the engineers perform a "parametric design" whereby the value of every parameter is recalculated to get the new values that will produce the larger SUV. A new SUV variant (species) is produced, but for all that, the body plan is identical. Changing the body plan would be orders of magnitude more complicated, involving a far greater number of simultaneous changes to yield a functional vehicle. This illustrates why microevolution cannot change body plans.

Dog breeding also shows the power and limitation of microevolution. Dogs can vary enormously in size, shape, color, patterning, and voice. The process of variation in this case is helped along by artificial selection—breeders aiming for some particular result many generations away. This is something that natural selection, with its what-have-you-done-for-me-lately obsession, cannot do. But even with the benefit of dog-breeder foresight and planning, dogs do not gain new fundamental design parameters. It is only the value of the parameters that is changed through the breeding process. A dog breeder cannot change a dog into a rabbit by breeding, because rabbits have different fundamental parameters in their skeleton, muscles, and organs.

The problem is that it takes many precise, simultaneous changes to produce any meaningful new structures or systems. For example, to evolve a closed circulatory system for blood from an ancestral open circulatory system would require numerous simultaneous changes. The same goes for changing one animal body plan into a distinctly different animal body plan. Try making the changes a bit at a time, a generation at a time, and we quickly find ourselves with a dysfunctional animal—an evolutionary dead-end. What's required is a huge number of simultaneous changes, but this is not possible through a process of random evolutionary accidents.

Peppered moths are often cited in textbooks as proof of evolution in action. The darker variety of these moths are said to have become dominant when, after soot from industry darkened tree trunks, being dark allowed the moths to camouflage themselves on soot-darkened tree trunks, the better to elude predators. But this is a very modest change. Also, the dark variety already existed and merely became more numerous.[31]

Four Constraints on the Evolutionary Mechanism

In the introduction I mentioned that evolution is a highly constrained process. Here I expand on this point. There are at least four major constraints on the evolutionary process:

Constraint 1: The Limitations of Incremental Change

Evolution can only generate novel forms that can be built up in small, incremental steps where each step involves a fully functioning design. The steps must be small because chance mutations become absurdly improbable where they involve lots of simultaneous, fortuitous genetic alterations in a single go. You can randomly change a letter or two in a sentence and occasionally get a new coherent sentence. For instance, in the previous sentence the "g" in "get," if it randomly mutated to an "n" would "net a new coherent sentence"—a bit slangy but still a coherent sentence slightly different from the original. But randomly change numerous letters in a lengthy sentence simultaneously, and do this sixty times a minute, and the sun will go red giant before you get a fundamentally novel coherent sentence.

Darwin's clever idea was proposing a way to stack up plausibly small changes. The challenge for his mechanism is that each generation has to be functional. After all, a stillborn organism doesn't reproduce and pass on its genes, no matter how clever some of its mutations might have been when combined with other clever mutations down the road. For that creature, there is no *down the road*. It's the end of the road. This unyielding stricture is why evolution can only generate a novel form where each small step in a long series of incremental steps results in a fully functioning design.

I have taught engineering design to thousands of students over the last thirty years at world-leading institutions. When I teach students how to design, I tell them they must not restrict themselves to step-by-step change, because that would exclude the vast majority of design possibilities and also the most innovative ones. But this is the lot of the evolutionary process. It does not have access to the vast majority of design solutions, including the most innovative ones.

Constraint 2: The Limitation of Survival Needs

A second constraint on evolution by natural selection is that it can create precious little beyond what is necessary for survival, at least nothing functional and of any sophistication. (We will consider sexual selection and the idea of evolutionary spandrels later.) This principle has been stated by the prominent evolutionist Steve Jones: "Evolution does its job as well as it needs to, and no more."[32]

This constraint has also been acknowledged by Alice Roberts, who says evolution only evolves structures that are "good enough" for survival and reproduction, and not well beyond that.[33]

When I teach students how to design, I tell them not to restrict themselves to the necessary functions. I advise them to add functionality and beauty beyond what is strictly necessary to fulfil the basic functions. For example, I tell them that a car should not just get from A to B but should be comfortable, exciting, and beautiful as well. In contrast, evolution cannot add extra function merely because doing so would be cool, or add beauty for the sake of beauty. Darwin, recognizing this about his theory, and further recognizing that the living world is bustling with examples of great beauty and seemingly excessive function, offered various explanatory patches. But as we have seen and will see, those patches have their own problems and can take a creature only so far.

Constraint 3: Vestigial Structures

A third constraint on the evolutionary process is an inability to reliably offload vestigial structures. According to evolutionary theory, almost any organism should contain structures from past ancestry that no longer have any use. The evolutionist Nathan Lents has explained that evolution is poor at deleting structures like bones, so the evolutionary paradigm predicts that a biological system like the human body should contain many useless parts.[34] The problem for evolutionists is that there is a growing catalog of "vestigial organs" that have turned out to have previously overlooked functions.

Constraint 4: Time

A fourth constraint on evolution is the time available to change one structure into another. The evolutionary process is an exceedingly slow one, so slow that even granting it millions of years, the process is expected to often produce half-baked results. For example, evolutionary paleontologist Jeremy DeSilva stated that evolutionary theory anticipates bad design in the human foot because there was not enough time to evolve an ape foot, optimized for tree-climbing and knuckle-walking, into a foot properly optimized for the bipedal locomotion of human beings.[35]

Devolution

Bacteria that become resistant to antibiotics are another commonly cited example of evolution via random mutations. Again, this is micro-evolution of a very modest sort. In many such cases, a minority of the bacteria in a population already had the genes for resisting the antibiotics, and in the presence of the antibiotic those bacteria, unsurprisingly, proliferated and came to dominate the bacterial population. In some cases, antibiotic resistance may have been the result of a genuinely new mutation, but the scientific literature shows that antibiotic resistance typically leads to a "fitness cost,"[36] where information is lost, causing degradation of a function. In these situations, antibiotic resistant bacteria get outcompeted by the normal bacteria when the antibiotic threat is removed. These are cases of devolution, degradation in design, not mutations with any prospect of leading to fundamentally new designs.

Such mutations lead to advantages for an organism in just the right circumstances, advantages that tend to become disadvantages when the organism is returned to the original environment. We see this in organisms large and small. The loss of color pigments in polar bears, descended from brown bears, is one instance of devolutionary change.[37] The polar bears blend into their arctic environment better, thanks to a devolutionary change.

In *The Edge of Evolution* and *Darwin Devolves*, biochemist Michael Behe describes how natural selection, in tandem with random mutations, commonly jettisons useful parts of sophisticated machinery in situations where that useful part is not relevant and where its loss lends the organism an advantage under very particular circumstances. As he puts it, "The First Rule of Adaptive Evolution: Break or blunt any functional gene whose loss would increase the number of a species's offspring."[38] He argues from experimental data and from mathematical analysis that such evolutionary solutions consistently crowd out the much more probabilistically difficult pathways required for the blind evolution of genuinely novel structures and body plans.

The principle of mutations tending to degrade a system is not surprising. In engineering it is well known that random changes in design tend to degrade performance or break the system. The same holds for software programs, which serve as a relevant parallel to the digital information we find in DNA. At best, random changes have a neutral effect or break something—leading to an advantage under particular circumstances, typically at the expense of versatility. Flightless birds on windy islands are one example.

Another example involves the *E. coli* bacteria in Richard Lenski's now-famous long-term evolution experiment. The bacteria in that study experienced various genetic mutations over the tens of thousands of generations that they were tracked in the lab, but they all turned out to be degradatory mutations—instances where something in the cell was blunted or broken. They weren't instances of the cell evolving some wholly new functional structure. In the most ballyhooed mutation, some of the *E. coli* developed the ability to eat citrate in the presence of oxygen. (It already possessed the ability to eat citrate outside the presence of oxygen.) At first blush it seemed that Lenski and his team had finally hit paydirt. But when they were able to determine how the new ability had been acquired, it turned out to be a case where "no new genetic information (novel gene function) evolved,"[39] and where traits were broken at the biochemical level.

So this long-term experiment—now more than thirty years and tens of thousands of generations of *E. coli* long—managed something noteworthy, but not what Lenski, a committed evolutionist, appears to have wanted to note: It demonstrated Behe's thesis that the joint evolutionary mechanism of random genetic mutations and natural selection is extremely limited in what it can achieve and its most impressive successes involve the creation of niche advantages through blunting or breaking.[40] In such cases, the evolutionary process does not build anything genuinely new, and in fact requires the breaking of molecular features.

Think of a car where you discard the backseat and the spare tire. Lighter car, better gas mileage, better handling. But such a process, as Behe notes, does not lend itself to building complex, functional new systems.[41]

More and more, investigators are finding that the modest adaptive triumphs highlighted in biology textbooks turn out, on closer inspection, to be examples of devolutionary mutations. Something in the organism's DNA broke, but broke in a way as to provide an advantage in some narrow circumstance.

It's unsurprising that the science media do not focus much on this finding. What committed evolutionist would be eager to contact the media to do a story recategorizing their favorite examples of "evolution in action" as cases of devolution? After all, evolutionists need to show that the evolutionary process created Earth's vast panoply of life, and for that, examples of biological devolution are of little help.[42]

Natural Selection: Conservator, Not Innovator

One might object that the above analysis has given too little attention to the role of natural selection. Let's remedy that deficit here. The first thing to note is that natural selection itself doesn't innovate; it only selects from among the innovations provided by random mutations. And as every engineer knows, coming up with innovations is the hard part. It's the thing in the history of life most in need of explanation. Also, if natural selection was such a big help, we would see it working wonders on the random mutations in contexts such as Lenski's long-term evolution experiment with *E.coli*, or in any of various natural experiments involving large populations and rapid generational turn-over, as with the malaria parasite. We don't.[43] Moreover, devolution is itself driven by natural selection.

Natural selection is a mixed bag. It does tend to weed out deleterious mutations, and can slow down the rate of devolution. However, it is blind to future needs and might weed out mutations that are presently deleterious but, when combined with a host of other very specific mutations, would provide a useful function down the road if allowed to remain. Thus, it cannot select for complex traits that require many mutations to be present before giving some advantage. This is why even Jerry Coyne admitted, "It is indeed true that natural selection cannot build any feature in which intermediate steps do not confer a net benefit on the organism."[44] Natural selection is all about the survival of the fittest *now*. It's too impatient, too short-sighted, to nurture the arrival of a fundamentally new design that remains several generations away. And if it cannot be expected to generate fundamentally new designs due to its short-sightedness, new designs of any sort, how much more does it lack the ability to fashion the many instances of ultimate engineering explored in these pages?

As noted in the introduction, many types of adaptation are not random but involve pre-programmed adaptability. For example, studies indicate a complex epigenetic mechanism causes Mexican cave fish to lose eyesight. Such discoveries further emphasize the limited role of random mutation and natural selection in adaptation.[45]

Contrary Evidence from the Fossil Record

Darwin was aware that the fossil record did not provide good evidence for macroevolution. The Cambrian explosion is the most famous example of the sudden appearance of complex organisms in the geological record. This event involved a sudden radiation of novel animal body plans, including practically all major animal phyla. Before the Cambrian, most organisms were relatively simple, individual cells or small multicellular organisms. Darwin hoped that future fossil discoveries would mitigate the challenge of the Cambrian, but discoveries since his time have only exacerbated it.

Or consider flowering plants. Darwin famously stated that their sudden appearance in the fossil record was perplexing on evolutionary grounds. As he put it, "The rapid development, as far as we can judge, of all the higher plants [flowering plants] within recent geological times is an abominable mystery."[46]

Today it remains a problem for evolutionary theory. Flowering plants have a very complex design, and yet they suddenly appear in the geological record (in the early Cretaceous period) with no fossil precursors.[47]

As far as the fossil record goes, the main difference between Darwin's time and today is that there are now many more "abominable mysteries"—that is, we know of many more biological forms that suddenly appear in the fossil record with no evolutionary history.

An example is flying insects like dragonflies. Paleoentomologist Günter Bechly noted that what he described as the "Carboniferous Explosion of Winged Insects" presents another serious challenge to evolutionary theory.[48] They appear suddenly in the fossil record, without credible near-precursors, and their fossils reveal how ancient dragonflies were very similar in design to modern dragonflies, meaning they were already enormously complex when they made their first appearance in the geological column.

I know dragonfly flight is immensely complex, because I had a major grant from the UK Ministry of Defense to study their flight and develop micro air-vehicles that copied their design features.[49] Dragonflies have a design more advanced than that of an F16 fighter jet. For a dragonfly to evolve from some simpler insect form would require hundreds if not thousands of step-changes in design. Yet dragonflies suddenly appear fully formed in the fossil record.

The only claimed precursors to dragonflies are Palaeodictyoptera. But these insects were fully optimized for flight, like dragonflies. In fact, in one sense Palaeodictyoptera were even more complex because they had three pairs of wings, one on each thoracic segment. And these also appear abruptly in the fossil record. "The oldest fossil winged insects belong to the orders Palaeodictyoptera (e.g., *Delitzschala*) and to the giant dragonfly order Meganisoptera," Bechly noted; "thus they were already equipped with the complete wing apparatus. There is not a single transitional form."[50]

Indeed, the fossil record shows dragonflies (as complex as modern dragonflies) appearing soon after flightless insects, and flightless insects as a group themselves appearing suddenly. Thus, in one major review on the origin of insect flight, David Alexander laments, "Unfortunately, the fossil record has so far offered little help in understanding how insect flight arose." As he emphasizes in a bullet point: "The incomplete fossil record is little help: there is no Archaeopteryx for insects."[51]

The fossil evidence supports a model wherein life was designed and generally instantiated at approximately the level of families, as for instance, the dog family (Canidae—e.g., wolves, coyotes, jackals, dogs) and the cat family (Felidae—e.g., tigers, lions, cheetahs, house cats). Each of these families appears to have frequently diversified into very similar species within each family, but this is a decidedly non-Darwinian understanding of life's history. This model can be thought of as an orchard of trees as opposed to Darwin's picture of a single "tree of life." In this "orchard of life," each tree represents a family of organisms, with each family starting from a single trunk and branching out into the many species of that family.

Evolution: The Magic Wand

If you look in a college biology book for direct evidence for evolution, as opposed to imaginative inferences and just-so stories, it does not give examples of macroevolution. Instead, it gives modest examples of small scale adaptation (microevolution) such as Darwin's finches, peppered moths, cave fish losing their sight, and bacteria developing antibiotic resistance. The fact that textbooks routinely offer up examples of microevolution as key evidence for the grand microbe-to-man evolutionary story illustrates the paucity of direct evidence for macroevolution in action, since you can be sure that if any such existed, evolutionists wouldn't be trumpeting the fact of finch beaks modestly increasing and decreasing in size.

I once asked a very senior biologist (a head of school at a top university who has hundreds of scientific papers to his credit) if he realized that evolutionary theory was a worldview that involved un-proveable faith commitments. He replied, "Of course evolution is a worldview." He was not at all religious but was adamant that evolutionary theory was based on the unprovable assumption that there is no intelligent designer responsible for the things of nature.

Imagine a court case where the judge says the jury should ignore any contrary evidence and find the defendant guilty. This is what happens today in origins biology.

No one disputes that biological systems look designed. Even atheist Richard Dawkins admits this. As he put it, "Biology is the study of complicated things that give the appearance of having been designed for a purpose."[52] Biological forms look designed, but we are told to ignore this. And if design is ruled out of court as an explanatory option, then—so goes the reasoning—some version of evolutionary gradualism must be true.

I asked the senior biology researcher mentioned above what he thought about such thinking. Evolution, he said, is like a magic wand. You wave the magic wand of evolution and say, "Evolution did it." Such magical thinking holds sway over a large swath of scientists focused on biological origins, but a sea change may be in the offing.

Not everyone is willing to pretend the naked emperor is dressed in royal robes. There is a list of Darwin dissenters published by the Discovery Institute, which now includes over one thousand PhD scientists, several of them at elite academic institutions. And the list, keep in mind, represents only those willing to come forth publicly with their doubts about Darwinism at risk to their careers. From my numerous firsthand conversations with life scientists at leading academic institutions, the doubts extend far beyond this group of dissenters willing to put their names on paper. And for more than a few of these, the doubts about modern evolutionary theory are accompanied by a growing openness to a fundamentally different explanation.

18. Intelligent Design Ascending

Human ingenuity… will never devise any inventions
more beautiful, nor more simple, nor more to the
purpose than Nature does; because in her inventions
nothing is wanting, and nothing is superfluous.
—Leonardo da Vinci[1]

Growing Support for ID

Before Charles Darwin or any other student of the biosphere had done any theorizing about biological origins, it was obvious that if one were to provide an adequate explanation for the origin of the living world, one had to account for both the marvels of the biological realm and for grim realities such as disease, deformity, and death. Thus, any theory of origins worth its salt, then or now, will offer some account for this full range of phenomena—beauty and ugliness, function and dysfunction, shadow and light.

Both modern evolutionary theory and the theory of intelligent design attempt to do so. The question is: Which offers a more satisfactory account? That is an evidential question, not a matter of taking a poll. But I will say that over the last thirty years I have seen strong support for intelligent design in academia, and the support is growing. Even among academics who remain on the sidelines, an increasing number are now aware of the mounting evidence for intelligent design.

It is hard to overstate how different the reality is from the surface-level confidence projected by the public relations wing of modern evolutionary theory. The vast majority of Darwin dissenters have yet to go public. I know this from direct personal testimony. If all of them went public tomorrow, it would create a tidal wave of dissent against the seemingly impregnable paradigm of evolutionary materialism.

Additionally, several senior biologists who are not proponents of intelligent design theory have freely admitted that intelligent design increasingly is the proverbial elephant in the room. When researchers are investigating the intricacy of molecular machines in the cell, everyone is thinking, "That really, really looks designed!" But they are not allowed to express this thought because intelligent design is deemed verboten.

Cracks, however, are appearing in the reigning paradigm. In a 2016 meeting of the Royal Society, the oldest and arguably most prestigious scientific organization in the world, top-level evolutionary biologists got together to consider whether neo-Darwinism was bankrupt and to push forward looking for a replacement. There also have been various high-level defections from Darwinism to design (e.g., the distinguished German paleontologist Günter Bechly and the famous atheist philosopher Antony Flew).

Additionally, there are now hundreds of peer-reviewed scientific papers supportive of intelligent design,[2] a very different picture from a generation ago when Darwinists regularly insisted that intelligent design couldn't be considered a scientific theory because it hadn't made it into the peer-reviewed literature.

Homology of the Vertebrate Limb

One of these many peer-reviewed papers is mine, published in August 2024 in a leading journal. The article directly challenges an iconic argument for evolutionary theory, Darwin's homology argument.[3] Figure 18.1 shows the iconic diagram in support of that argument— the similar skeletal layouts of the pentadactyl limbs of six vertebrate species. Variations on this diagram are shown to students around the world, with the similarity of design across the different limbs presented as proof of evolution.

Human arm	Whale flipper	Bird wing	Human leg	Cat hindlimb	Frog hindlimb

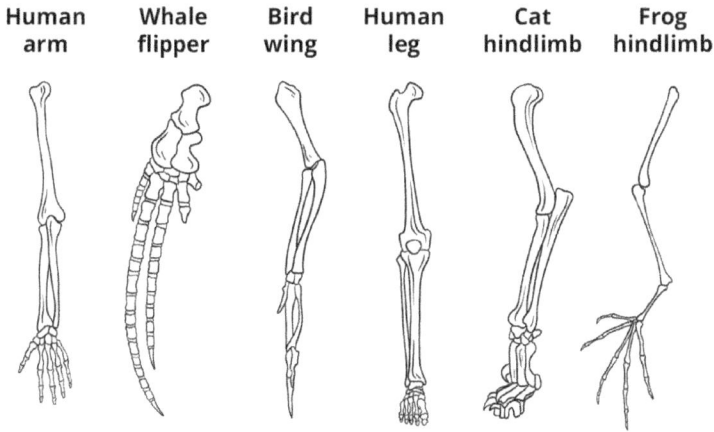

Figure 18.1. Vertebrate limb pattern.

Darwin assumed that the vertebrate limb pattern was *not* universally optimal for all these animals, and from this he reasoned that the similarity of design could be explained only by evolutionary inheritance, since a wise designer, such as God, would surely have come up with optimal designs from scratch rather than going back to the same basic design architecture over and over again and settling for the suboptimal. In particular, Darwin assumed that the whale flipper had no need for elbow, wrist, and finger joints and that these features showed that the whale had come from an ancestor that previously walked on land. This thinking has been repeated ever since in high school and college biology textbooks.

Over the last thirty years, while researching animal biomechanics, I became convinced that Darwin was mistaken on this point and that the limb pattern is in fact optimal for all the animals that possess it (much as the wheel or pulley is optimal in a wide variety of engineering contexts) and thus is well explained on the grounds of intelligent design. I collected all my findings into an article where I applied engineering insights from areas such as hydrodynamics and aerodynamics to explain how the vertebrate limb is indeed universally optimal. The result was a 15,000-word paper with some fifty diagrams.

In the case of the whale flipper, I showed that the whale's fingers and wrist are essential for controlling the shape and stiffness of the flipper. For example, when a whale slows down, the flippers would deform in shape (due to the huge braking forces) unless the muscles of the fingers and wrist resist the external loads. It is important to maintain the shape of a flipper during swimming in order to maintain a low hydrodynamic drag coefficient; otherwise swimming will become inefficient. The fingers and wrist are also needed to fine-tune changes in flipper shape. For example, when the tips of flippers are curved, this can reduce tip vortices, swirling patterns of water that increase drag.

I also referenced research studies showing that the whale's fingers each have muscles for flexion and extension, just like human fingers, and that these have been observed to be fully functional in whales. A similar situation is found in the design of bird wings. The wrist and finger joints of birds are essential for fine-tuning the shape and stiffness of the wings for efficient and agile flight.

In summary, whale flippers and bird wings are not stiff paddles. They are actually structures that can morph in shape and stiffness in order to maintain high efficiency during swimming and flying, and the pentadactyl design is key to how well they perform.

It is helpful to point out that most muscle activity in flippers and wings involves isometric muscle force, where muscles are used to prevent movement. This contrasts with land mammals, where most muscle movement is isotonic muscle force, where muscles are used to cause movement. This means that it is not so obvious that flippers and wings require joints and muscles; and yet the reality is that joints and muscles are just as important in flippers and wings as they are in arms and legs.

Charles Darwin was not aware of these hydrodynamic and aerodynamic considerations. He simply assumed that the flippers contained vestigial structures or structures derived from common ancestry rather than designs chosen from the ground up for the purpose at hand. Has anything changed? More than a century and a half later, it is common to read on museum websites that the whale flipper is a rigid paddle, implying that the fingers and wrist are vestigial.[4] But my paper showed how this is mistaken. It was titled "Universal Optimum Design of the Vertebrate Limb Pattern and Lessons for Bioinspired

Design." I adopted this title to make it abundantly clear that I was challenging Darwin's homology argument. The paper appeared in *Bioinspiration and Biomimetics*, a world-leading bioengineering journal. In the conclusion I explicitly challenged modern evolutionary theory:

> The great versatility of the vertebrate limb pattern challenges the limb homology argument that the skeletal layouts of the whale flipper and bird wing are not what would be expected for those applications and make sense only when seen to be a consequence of evolutionary inheritance. This paper argues that the vertebrate limb pattern is so versatile that it is actually highly optimal not just for arms and legs but also for flippers and wings. All the musculo-skeletal structures of flippers and wings are actually fully functional and fully explainable in terms of optimal design.[5]

The paper, although novel from an evolutionary perspective, was not voicing a view that the bioengineering community regarded as scandalous and beyond the pale. Support for the paper is evidenced by the fact that it was not only accepted but given a category of "Review Paper," meaning it was deemed an important contribution describing the state-of-the-art for the subject. In the first twelve months of publication it was downloaded more than 3,500 times, and cited six times by other professional papers. Those figures are high after just one year of publication.

Engineers are very pragmatic. If the vertebrate limb is universally optimal, then engineers want to know this, because they want to know if they should be copying limbs to make better robotics. My paper made it clear that copying the limb design of all kinds of animals could prove immensely fruitful. The fact that I was criticizing Darwinism was deemed an insufficient reason to stop the publication of a paper that would help the robotics community.

Unfortunately, the paper's discoveries, and the logic undergirding them, have not made front-page headlines. When discoveries are made that contradict the naturalistic worldview, they rarely make it into the mainstream media or educational institutions. When students are shown this series of limbs in Figure 18.1, what they are rarely if ever told is that common descent is not the only logical explanation for common design patterns, and that a common designer reusing a versatile design template well explains the pattern. In the case of the vertebrate limb, such a conclusion makes particularly good sense because the various instantiations of the pattern represent optimal design.

Irreducible Complexity

Irreducible complexity is a clear scientific test of whether a system could have evolved step by tiny step. Indeed, Darwin himself famously stated that his theory would collapse if a biological structure could not evolve in this way, with each intermediate structure having a useful function. As he put it, "If it could be demonstrated that any complex organ existed, which could not possibly have been formed by numerous, successive, slight modifications, my theory would absolutely break down."[6] Darwin was basically inviting people to apply the test of irreducible complexity to his theory.

The term "irreducible complexity" is used by biochemist Michael Behe to describe any biological structure that possesses many indispensable parts, parts without which the system fails.[7] This means that the probability that the structure is built one tiny step at a time over many generations is no better than raw chance, since the number of mutant organisms won't be increased by natural selection if an intermediate structure provides no functional advantage. The Darwinian mechanism is rendered impotent.

It's not difficult to find biological machines and systems that appear to be irreducibly complex. Just the opposite. The challenge is finding any sophisticated biological machine or system that does *not* appear irreducibly complex.

Evolutionists have attempted to offer various clever workarounds to this challenge, their favorite being the idea of co-option, in which the evolving biological entity co-opts a biological part or machine that was being used for some other purpose in the biosphere. When rebutting him, Behe's opponents often attack a strawman by claiming that he ignores this clever co-option solution. Not so, and his responses are worth reading in full.[8]

Here we boil those responses down to a few short paragraphs. First a bit of background on Behe's most famous biological illustration of irreducible complexity. He argues that the microscopic bacterial flagellum motor is irreducibly complex, needing numerous parts to function. Remove one of these thirty or so essential parts, and the flagellum is a bust. But critics note that the type III secretory system has several parts resembling those of the bacterial flagellum, parts that may have been co-opted on the way to evolving a bacterial flagellum. However, it turns out that many other flagellar parts appear unique and are not co-optable. Moreover, the type III secretory system seems to post-date, not pre-date, the flagellum and cannot serve as an evolutionary precursor to it.[9]

In engineering it is well known that common parts are used in many different subsystems, so it is no surprise from a design framework that parts of the flagellum could also be found elsewhere. Also, just because co-option could occur in principle doesn't mean it can explain the origin of any particular biological machine or system. Appealing to co-option just moves the problem to another place: How did the co-opted system evolve in the first place? One needs to provide actual evidence of a detailed evolutionary pathway, with or without co-option as part of the itinerary.

Further, in our uniform experience, co-option itself requires an intelligent agent to select, tailor, and fit the co-opted system into the emerging new design in order for the co-opted structure to fulfill its new role in the larger emerging system. Speaking the incantatory words "evolution" and "co-option" does not give us fundamentally new biological machines and systems any more than saying "abracadabra" and waving a magic wand would.

However, by testing and attempting to falsify Behe's argument against evolutionary theory and for intelligent design, these evolutionists implicitly acknowledge that his argument is testable/falsifiable in principle; and testability/falsifiability, as many an evolutionist will tell you, is a prize marker of a properly scientific hypothesis.[10]

ID critics counter this point by insisting that irreducible complexity is merely a negative argument against evolutionary theory, not a positive argument for intelligent design. But that's another strawman characterization of Behe's argument. As Behe explains, any time we can trace an irreducibly complex system back to an originator (not to a story about an originator, and not to an educated guess about the originator, but to the actual originator—e.g., the incandescent light bulb to Thomas Edison, the printing press to Johannes Gutenberg), we always come to an intelligent agent. This uniform and repeated experience encourages us to identify intelligent design as the best explanation for irreducibly complex biological machines, all the more so when this positive evidence is yoked to the negative evidence telling us that mindless evolution lacks the causal power to produce such machines.

This conclusion from our uniform experience, Behe further notes, is corroborated by straightforward reasoning about the causal powers of designing minds. Intelligent agents possess foresight and the capacity to pursue distant goals, something that chance and brute natural forces lack. An intelligent agent doesn't need the invention to emerge one tiny random step at a time, with each step involving a functional offspring able to survive and reproduce. An intelligent agent can skip over a whole series of dysfunctional intermediate steps as the agent brings together various parts into an irreducibly complex whole.

Irreducibly complex machines and systems bespeak conscious design from the fact that such machines and systems involve, as Behe puts it, "a purposeful arrangement of parts." After all, only conscious agents have purposes, and only conscious agents have the capacity for foresight and planning to pursue those purposes.

This latter idea has been around for a long time. The ancient Roman philosopher Cicero, in making an argument to design in nature, described how a water-clock bears marks of design because its many parts work together for a useful purpose.[11] And in 1802, the Cambridge University scholar William Paley published *Natural Theology*, in which he described how a precision mechanical clock bespeaks intelligent design due to how the many parts cooperate for a precise purpose.[12] He applied this observation to living things, arguing that the purposeful arrangement of parts in living things points to a mindful designer.

Behe built on this tradition by rigorously defining and illustrating the concept of irreducible complexity and by diving into what was unavailable to Cicero and Paley, what Behe memorably dubbed "Darwin's Black Box"—that is, the molecular biological realm of ingeniously engineered systems and machines that only became accessible to scientists in the twentieth century, well after Darwin's death.

Behe has shown how many molecular biological machines and systems exhibit high levels of irreducible complexity.[13] The revolution advanced by ongoing progress in microscopy and other technology has uncovered many new examples of irreducible complexity, from the bacterial flagellum that Behe made famous to the blood clotting cascade (also spotlighted by Behe) and many other molecular biological marvels.

The examples come so thick and fast at the microscopic level that it's easy to ignore what we are also uncovering at the visible, anatomical scale. I have published many papers describing the intricate complexity of biomechanical systems in biology. Several of these papers make it clear that anatomy is brimming with cases of irreducible complexity.[14] Many such examples have been given in this book. Additionally, a recent book by systems engineer Steve Laufmann and physician Howard Glicksman provides several clear examples of irreducible complexity in body-level systems, such as homeostasis and those involved in embryological development.[15] Such systems require complex sensors and feedback loops that have to be fine-tuned for the system to function, systems that would utterly collapse if any of the many key parts weren't in place. That is the wrong job for a mindless process without foresight, creativity, or the capacity to plan. But it is just the job for a cause with all of the above—namely, a designing intelligence.

Irreducible complexity is one of the biggest challenges to evolutionary theory today. I highlighted this in my 2024 review paper in the journal *Biomimetics* on multifunctioning in nature:

> Complexity in biological systems is sometimes labelled as an emerging property. However, it is very difficult to explain how a multifunctioning system could emerge from an initially single-functioning system because the first single-functioning system would have to be one of the very few solutions that could lead

to a later multifunctioning system. When discussing the origin of mechanical linkage mechanisms in animal joints, Muller has stated that it is very difficult to see how complex mechanical linkage systems can be developed in a bottom-up step-by-step process. Therefore, multifunctioning in biological systems such as limb joints presents a major challenge of irreducible complexity for evolutionary biologists.[16]

The Information Challenge

The information in living things also poses a major problem for modern evolutionary theory and strongly suggests intelligent design. Stephen Meyer has explored how biology is full of information-rich systems that include data storage, replication, processing, and quality control. And he argues that such information-rich systems are a hallmark of intelligent design.

Meyer demonstrates, moreover, that the first form of life on Earth had to have been an information-rich system with a fully functional, self-replicating, protein-synthesizing system of DNA, RNA, and proteins.[17] His arguments are corroborated by other lines of evidence we looked at in the previous chapter, including laboratory attempts to engineer the minimal viable cell (an entity that has proven to be enormously complex) and insights gleaned from ongoing efforts to construct a truly self-reproducing machine (such as we find in even the simplest cells), efforts that drive home with unparalleled force just how sophisticated any such entity must be to pull off the extraordinary trick of self-reproduction. Life itself, it would seem, is irreducibly complex.

Andy McIntosh, a professor of thermodynamics in the UK, has spotlighted another challenge facing the idea of blind evolution and pointing to intelligent design. He has shown how some chemical reactions—called endergonic reactions—are not spontaneous and therefore require not just a source of energy but also a "machine" of some kind to move from an equilibrium state to a non-equilibrium state.[18] In technical terms, an endergonic reaction has to raise the "free energy" within a system. An example of an endergonic reaction is photosynthesis, which changes low-grade energy (sunlight) into high-grade energy (sugar). This change can never happen spontaneously but needs a complex machine alongside the energy to create the highly ordered molecular structure of sugar.

It is often claimed that since the earth is an open system with an external energy source from the sun, the earth can therefore increase in order (decrease in entropy). However, energy on its own does not lead to increasing order (raising the free energy). In fact, energy on its own can only be converted to heat. Therefore, the sun on its own cannot cause endergonic reactions. It therefore follows that the sun's energy is not enough to cause abiogenesis.[19] The introduction of irreducibly complex machines is also required, and as suggested above, for their origin we have abundant negative evidence pointing away from mindless evolution and abundant positive evidence pointing toward intelligent agency.

Smart Adaptation: Another Elephant in the Room

Adaptation provides inadequate evidence for macroevolution, because it primarily involves either the shuffling of pre-existing alleles or epigenetic changes that don't involve actual changes in the DNA sequence. Even though random shuffling of this sort can result in increased fitness for survival, the changes that take place are limited to the information in the genes of the parents.

When considering adaptation, account also must be taken of the possibility that adaptation is not always based on random genetic changes. There is increasing evidence that organisms are sometimes pre-programmed with designs ready for new environments. A pre-programmed design means that there is genetic information already in existence ready to be deployed when the environmental demand arises. In such cases there is no need for new information to be created, because it already exists.

A paper in the journal *Nature* describes how some organisms can actively respond to an environmental shift by generating traits well suited to the altered context. Two instances: Cichlid fishes change the shape of their jaws when food sources change; leaf-mimicking insects born in the wet season are green, while those born in the dry season are brown.[20] More and more, scientists are acknowledging that adaptation often involves just these sorts of pre-programmed directed changes, what is often called epigenetic changes or phenotypic plasticity. This challenges the long-held view that adaptation is necessarily driven by random genetic changes.

Gene variation, it's important to note, has important purposes apart from helping a creature survive changes in its environment. For example, gene variation creates variety, which can have important functional roles, such as creating different facial designs, as with humans. Gene variation also creates beautiful variety, as shown in the diverse patterns of butterflies and flowers.

Gene variation also helps humans breed animals and cultivate plants for optimal use. The usefulness of design variation can be seen with human products like cars. Cars are deliberately designed to vary in size, color, and functionality in order to satisfy the needs of the market and help the manufacturer profit and survive.

Gene variation makes good sense from an intelligent design perspective, because it has so many design functions. One could argue the same from the Darwinian perspective, but a tie-breaker between these competing hypotheses (Darwinism and intelligent design) is that we are finding more and more evidence of directed adaptation, sometimes involving gene variation and sometimes epigenetic changes or other processes (design details that bespeak masterful engineering), even as we are also finding increasing evidence of the severe limits of what these undirected natural processes can achieve in terms of adaptive change.

Beauty for Beauty's Sake: Another Elephant in the Room

John Maynard Smith, an eminent theoretical evolutionist, a Fellow of the Royal Society, and recipient of the Darwin Medal, identified beauty as one of the biggest challenges for the theory of evolution, writing that "no topic in evolutionary biology has presented greater difficulties for theorists."[21]

The reason beauty is a problem for evolution can be illustrated from human architectural design. The embellishments on buildings like Buckingham Palace, such as the carved patterns on the pillars, have no other function than creating beauty. They do not make parts stronger or stiffer or cheaper to make. In fact, they are more expensive to make and require a lot of precise information. The aesthetic features therefore point to a designer who wants to add aesthetic beauty. In the same way, if there is beauty for beauty's sake in nature, that is powerful evidence for intelligent design.

Darwin himself agreed that if there is beauty for beauty's sake in nature, this would be evidence against his theory. "Some naturalists... believe that many structures have been created for the sake of beauty, to delight man or the Creator.... Such doctrines, if true, would be absolutely fatal to my theory. I fully admit that many structures are now of no direct use to their possessors, and may never have been of any use to their progenitors; but this does not prove that they were formed solely for beauty or variety."[22]

Here Darwin is showing commendable honesty in admitting that beauty could prove fatal to his theory. He also once confessed to his friend Asa Gray, "The sight of a feather in a peacock's tail, whenever I gaze at it, makes me sick!"[23] Darwin developed his complementary theory of sexual selection to account for such examples of superfluous beauty in the natural world, an idea he developed in *The Descent of Man*.

The idea, in a nutshell, is that a creature is inclined to choose the more beautiful sex partner, leading in some cases to the evolution of extremely showy ornamentation, as with various types of birds. But even Darwin seemed to realize that this patch on his theory of natural selection left something to be desired. In the same book, he admitted, "Many will declare that it is utterly incredible that a female bird should be able to appreciate fine shading and exquisite patterns."[24] Darwin is again showing thoughtful honesty here in conceding that, from within his evolutionary framework, it beggars belief that female birds could see and appreciate fine aesthetic details in the male.

Unsurprisingly, however, Darwin does not give up on his theory of sexual selection, an idea he likely came to see as a crucial buttress to his evolving theory of evolution. Instead he argues, "He who thinks that he can safely gauge the discrimination and taste of the lower animals may deny that the female Argus pheasant can appreciate such refined beauty; but he will then be compelled to admit that the extraordinary attitudes assumed by the male during the act of courtship, by which the wonderful beauty of his plumage is fully displayed, are purposeless; and this is a conclusion which I, for one, will never admit."[25]

Here Darwin is less honest, or at least less clear-thinking. The female Argus might appreciate the rooster's plumage and the verve of his strutting at only a very coarse level, too coarse to explain the origin of such exquisitely artful plumage; or, if at a refined level, we are left with the question of why evolution would develop in the female such extraordinarily refined and impractical tastes in male partners. The question is particularly pressing as regards pea hens and peacocks, the latter of whom are saddled with those enormous tails, hardly ideal for quick getaways when a predator arrives on the scene.

The evolutionist's answer is that perhaps the preference piggybacked on the peafowl's preference for bright ornamentation, which might have functioned as a proxy signal for male fitness, since only a robust peacock could get along despite being burdened by such unnecessarily large tail feathers. Bigger peacock tails offered more brightness, the pea fowl's delight, and then a feedback loop began. Give an evolutionist long enough and he will construct an imaginative story that saves the appearances.

The problem with this story is the peacock tail in its totality. It's not just big and bright. It's an astonishing work of art. This reality is impossible to quantify in reductionist terms. The move for the Darwinian reductionist is to do what reductionists do: reduce—in this case, by substituting in place of the work of great beauty some of its constituent aspects—here, size and brightness. But that is to explain away rather than explain. It's a bit like describing Shakespeare's tragedy *Hamlet* as a fairly long assemblage of words involving an unusually large vocabulary and several lovely turns of phrase. What's missing from this description is the overarching beauty and genius of the work. In the same way, the Darwinian reductionist's explanation for the peacock's stunningly beautiful tail explains only a poor substitute for that beautiful reality, not the work of art itself.

The idea that some creatures associate beauty with health, and then associate health with good genes for their offspring, has several problems. How can female birds recognize intricate details, such as borders and mathematical patterns, as beautiful? And why should a female animal associate beauty with health? And why should the female know that half the genes of her offspring will come from the male that she mates with? Female birds do not have a degree in genetics.

What about flowers? Are they beautiful merely because their beauty better attracts pollinating insects, a type of fitness the evolutionary process might naturally select for? Flowers exhibit an incredible range of beautiful features in terms of colors, patterns, textures and smells—far more than is necessary for attracting pollinators, given that insects can be attracted by simple colors and smells. And why should we find them so beautiful? They have no survival value for us. Darwinism has no ability to explain this.

A good theory should explain all the evidence, or at least as much of the evidence as possible. Here Darwinism fails while a design paradigm succeeds, since designing intelligences are known to create beauty for beauty's sake (in addition to creating it for other purposes as well).

Then there is the matter of musicality in birds. William Homan Thorpe, a world-leading expert in birdsong at the University of Cambridge, acknowledged it was difficult to see the selective advantage of the musicality of birds. "It is hard to imagine any selective reason for the extreme purity of some bird-notes," he commented, and "we do find a great deal of elaboration which goes beyond anything which would seem to be biologically advantageous."[26]

I agree. There are many aspects of birdsong that have no selective advantage. I wrote a chapter on the wonders of birdsong for my book *Hallmarks of Design*.[27] Some birds possess an uncanny degree of musicality, producing songs with a key signature, time signature, and musical structure such as phrases and themes. Songs have been found to have advanced musical features, such as finishing with finality[28] and transpositions from major keys to minor keys.[29] Some birds can even perform duets and quartets.[30]

In most cases, male songbirds use their singing not to attract a mate but to mark their territory and intimidate rivals.[31] This means the evolutionist has to argue that higher levels of musicality are more frightening.

Purposeful Overdesign

Another evidence for intelligent design is that of overdesign. If *Homo sapiens* arose from a mindless and gradual process of micromutations and survival of the fittest, we should be little better than just good enough for surviving and reproducing. Recall from Chapter 16 the summary of this idea by evolutionist and atheist Steve Jones: "Evolution does its job as well as it needs to, and no more."[32] A straightforward application of the theory leads to the expectation that we should not have skills and intelligence far beyond what is needed for survival and reproduction. Put simply, we should have brains and hands evolved only just enough to be skilled hunter-gatherers. But of course we are clearly capable of so much more.

One of the most obvious examples of overdesign is the human mind. It has an extraordinary capacity for memorizing information. One of the best known memory competitions involves memorizing the number pi, which is the circumference of a circle divided by its diameter. The reason pi is chosen is that it has no pattern to its series of digits, which go on to infinity. The first thirty digits of pi are:

3.141592653589793238462643383279

It takes quite a while to remember these thirty digits. However, with motivation it is possible to remember a lot more. The Guinness World Record for the feat is 70,000 digits, as recited by Rajveer Meena in 9 hours and 27 minutes in March of 2015.

Humans also have an amazing capacity for creativity. Mozart wrote over eight hundred musical compositions at a rate of almost one per week. Most of these compositions were highly complex and original works of art. Scientific pioneers such as Isaac Newton and James Clerk Maxwell were extremely creative and visionary in devising very complex mathematical formulae that illuminated previously hidden aspects of the natural order. There also have been engineers who have demonstrated great creativity in developing technology, inventors such as Leonardo da Vinci, George Stephenson, Isambard Brunel, and Thomas Edison.

And it is not just famous people. Ordinary humans have a capacity for creativity and deep thought that far exceeds what is needed for hunter-gatherer survival. But the capacity makes perfect sense if it is the result of a designing intelligence, with humans created in part to function as the stewards of the natural world.

One might counter that if this was the intention, then why have humans made such a mess of things with their greed, violence, and exploitation? This question draws us into the theological issue of the problem of evil. The Judeo-Christian tradition answers that question, but a detailed look at that answer takes us beyond the scope of this book. Suffice to say that humans have the demonstrated capacity to wisely steward the natural world but often choose not to do so. If the designer had chosen to engineer faultless automatons, without free will, presumably those automatons would have avoided our mistakes. But here we encounter the mother of all engineering trade-offs: In principle, a creation stewarded by perfectly engineered automatons instead of by free beings would make no mistakes, but it also would be incapable of making real choices and, into the bargain, performing acts of real courage, sacrifice, and love along with their darker opposites.

Evolutionists have long acknowledged that the overdesign evident in human intelligence is a challenge to the theory of evolution. Alice Roberts, for example, acknowledges that the human intellect appears to be overdesigned: "There has to be an evolutionary reason for growing that brain big…. Because it's a very demanding organ it wouldn't have grown bigger unless there was a really important advantage to us in having big brains and that has been something which has taxed paleoanthropologists for decades and decades and I think still taxes everyone in the field."[33] Our human capacity for language poses the same problem. Harvard evolutionary psychology professor and atheist

Steven Pinker frankly concedes that "human language is an embarrassment for evolutionary theory because it is vastly more powerful than one can account for in terms of selective fitness."[34]

Overdesign fits the design paradigm quite nicely, but Roberts, a committed evolutionary materialist, will not consider the design hypothesis.

Alfred Russel Wallace, the co-discoverer of the theory of evolution by natural selection, was more flexible on this point. As we saw in Chapter 16, he eventually concluded that natural selection could not explain the origin of the human intellect. But his break with his friend Darwin was actually even more far-reaching than this. "After forty years of further reflection," he determined "that not man alone, but the whole World of Life, in almost all its varied manifestations, leads us to the same conclusion—that to afford any rational explanation of its phenomena, we require to postulate the continuous action and guidance of higher intelligences; and further, that these have probably been working towards a single end, the development of intellectual, moral and spiritual beings"[35]—that is, the fashioning of creatures such as ourselves.

Wallace was a leading evolutionist of the nineteenth century and co-founder of the theory of evolution by natural selection, so his conclusion should not be dismissed lightly.

Human hands, as we saw previously, are another area where humans appear to be overdesigned. Our hands are clearly designed for extreme levels of skill in areas like farming, technology, music, art, craftwork, carpentry, and cooking. Such skills are far beyond what are useful for primitive hunter-gatherers, but such capacities are not surprising from a design framework, since it leaves one open to the possibility that humans were designed with all those capacities in order to be able to steward the natural world. After all, if you build a lawn mower, you build it with the capacity to cut grass. If you build a steward of the earth, you give that steward the ability to foresee, plan, cultivate, build intricate things with his hands, invent, and organize socially to pursue larger projects such as the construction and maintenance of irrigation canals, terraces, and orchards.

As discussed in Chapter 6, the human voice also appears overdesigned. It is capable of fast, complex, and nuanced communication as well as beautiful singing. Such skills are an awkward fit with evolutionary theory, with its model of bit-by-bit unguided evolution. Sexual selection is usually brought in here to save appearances—the singing voice evolving to attract mates; but the tens of thousands of engineering tweaks required to achieve such an evolutionary arc are left obscured by the gauzy light of the evolutionary just-so story. In contrast, an intelligent design framework faces no such difficulty, since it leaves open the possibility that humans were designed by a master creator to be stewards and sub-creators, creatures meant not only to survive and reproduce but also to enjoy life and flourish culturally.

Spandrels to the Rescue?

Evolutionists Stephen Jay Gould and Richard Lewontin have attempted to counter such examples of superfluous beauty and overdesign in the biological realm by appealing to the metaphor of decorated spandrels in St. Mark's Basilica in Venice. That is, they argue that some biological features are mere byproducts of some more functional feature. In the case of the basilica, the arches are the functional structures allowing for doors and windows, and the spandrels—the roughly triangular shapes between adjoining arches—just came along for the ride. In the same way, argue Gould and Lewontin, many biological features are "spandrels," mere unintended byproducts.[36]

Consider the idea as applied to the curiously advanced musicality of much birdsong. Natural selection would not select so much superfluous musicality. It's even a stretch to imagine that sexual selection would have done so. The bird's advanced musicality, rather, is a spandrel, Gould and Lewontin would argue, a byproduct of something else that did serve a function.

The problem with such explanations is twofold. For one, the appeal to a "spandrel" is often extraordinarily vague, handwaving without ever identifying a causally adequate means of building the exquisite feature in question—in this case, the advanced musical skill of some bird species. Additionally, the metaphor actually supports a design paradigm, for the spandrels of St. Mark's Basilica are beauti-fully ornate only because the creators of the basilica chose to make the extra effort to overlay the spandrels with beautiful designs. So a straightforward application of the metaphor would lead us to see in the superfluous beauty we find in birdsong, and in so many other corners of the living world, the exuberant creativity of a great artist.

Biological Wonders: The Gold Standard of Design

As this book has argued in the area of biomechanics, although there is evidence of degradation in the biological world—e.g., aging and harmful genetic mutations—we do not find canonical designs of par-ticular species that manifest shoddy or substandard engineering. Even the humblest of plants and animals are filled with engineering solu-tions that make our best human engineers look like bumbling children by comparison. When top scientists and engineers look at the design of even just a single cell, they have yet to fathom its layer-upon-layer of engineering sophistication, much less identify concrete avenues to improve it in a way that takes account of all the various tradeoffs. The same holds for when researchers study macro-level features such as skin, bone, and connective tissue.

This is why an entire subdiscipline of biology has grown up, bio-mimetics, where engineers try to copy nature to improve human tech-nology. Whether or not they admit it, these researchers, me included, are apprentices of a most intelligent, a master, engineer.

19. Ultimate Engineering Throughout Nature

A Theistic Design Paradigm Anticipates Ultimate Engineering

Many years ago I was asked to redesign the separation mechanism for the main stage of the Skylark sounding rocket, one of the most successful rockets of its kind used in Europe. Off-the-shelf engineering solutions wouldn't cut it. The task demanded ultimate engineering, and I had a big budget for the work, so I spared no expense of time or money to design the best possible solution using the best possible materials and machining methods.

When I look at nature, I see that every plant and animal contains ultimate engineering, and at levels of design beyond even the best rocket science. I also see creativity maximized in terms of diversity of designs and applications.

I'm far from alone in this assessment. We often hear scientists gushing over the masterful designs found in nature. In 2024, for instance, researchers published a paper with the following title: "Bio-Inspired Designs: Leveraging Biological Brilliance in Mechanical Engineering."[1] In every area of biology scientists are discovering brilliant engineering.

If the scientific theory of intelligent design is yoked to a theistic framework, then nature is approached with the assumption that the designer has perfect knowledge of scientific laws and materials and could and would employ such knowledge to achieve ultimate engineering. This expectation is being confirmed in innumerable ways.

How Ultimate Engineering Is Achieved

Creating a design that can achieve performance at the limit of what is possible in engineering and biological systems is difficult in the extreme. It can only be achieved by fine-tuning a design in several ways simultaneously. This includes employing the following:

- A masterful combination of physical effects.
- Materials with ideal properties.
- Ideal geometries and shapes.
- Unparalleled miniaturization of components where appropriate.
- Extreme multifunctioning of parts and subsystems.
- Highly integrated parts and subsystems.
- An "operating system" with high intelligence and control.

As we saw in the first part of this book, these strategies can be seen throughout the human body. The human arched foot ingeniously combines two very different physical effects: the elastic spring effect for shock absorption when landing and the high-stiffness physical effect for push-off. The materials of the foot, such as the ligaments, have ideal properties of toughness and strength to withstand the loading in the foot. The complex geometry of the femur bone at the knee joint allows for the rolling motion that leads to the large range of motion of the joint. The miniaturization of the sphincter muscles of the iris enables the shutter system of the eye to be extremely compact. The integration of the nervous system with the muscles and sensors of the body leads to extreme levels of skillful movement and sensing. And the high capacity of the human brain allows for creative thought and integrated control of the body's many subsystems.

Ultimate engineering is not confined to the human body. It can be observed throughout the biosphere. It is impossible to do a comprehensive study in one chapter. The following sections give just a few examples of designs in the living world at the limit of what is possible.

DNA: The Ultimate Information Storage System

DNA is a clear example of ultimate engineering. It's a long molecule located in each one of our living cells (with the exception of mature red blood cells) and contains information crucial to how a cell will grow, produce essential materials, and interact with other cells. DNA has a double-helical structure in a very long string, the backbone of which is made up of four different small subunits called bases. The bases are adenine (A), cytosine (C), guanine (G), and thymine (T). Each cell carries around six billion of these tiny bases in DNA molecules. These four bases can be thought of as the four-character alphabet of the DNA language.

Researchers marvel at the data storage capabilities of DNA, which may approach the limit of what is physically possible. For example, in 2021 *Scientific American* carried an article entitled "DNA: The Ultimate Data-Storage Solution."[2] The article explained how the double helix can archive a staggering amount of information in an almost inconceivably small volume. It is estimated that, in theory, 433 exabytes of data could be stored in just one gram of DNA. One would need around a billion 512GB flash drives to hold that much information.

Ultimate Materials

When we look beyond the human body, we find that the biological realm contains many extraordinary materials that are so fine-tuned for specialized functions that they appear to be at the limit of what is possible. In most cases, the designs are so intricate that engineers cannot fully mimic them.

Take spider silk. It is the toughest fiber known and epitomizes elegant perfection. Figure 19.1 gives a glimpse into its intricate structure. The silk consists of many strands of a protein fiber spun effortlessly by the spider. At the molecular level, each silk fiber is assembled by thousands of nanofibers that run parallel to each other. Each nanofiber is less than one thousandth of a centimeter thick.

Engineers are in awe of the performance of spider silk. The silk is around five times stronger than steel for a given weight and yet is highly flexible. It can function in extreme temperatures, from -40°C to 200°C. Some types of spider silk are tougher than Kevlar, the material used in bulletproof vests.

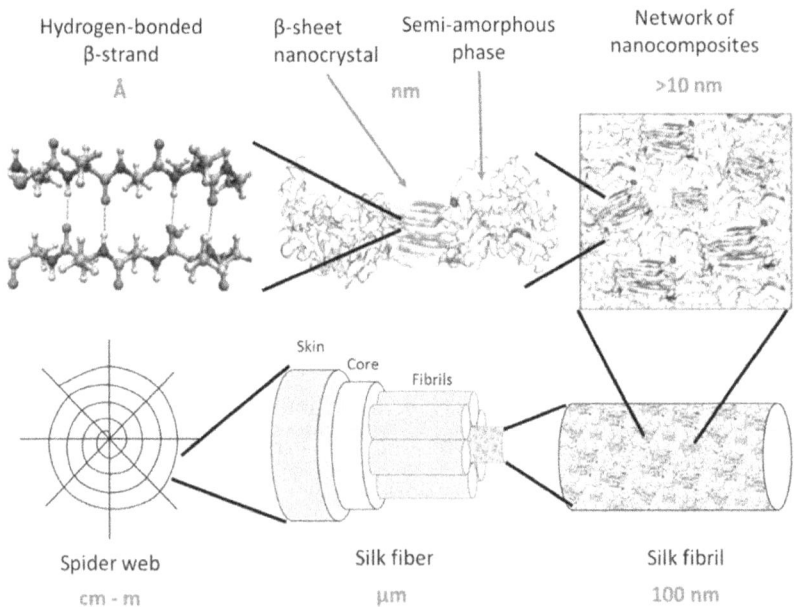

Figure 19.1. Hierarchical structure of spider silk.

A spider spins different types of silk for different parts of the web, such as silk optimized for strength versus silk optimized for elasticity. A spiderweb can instantly stop a relatively heavy flying insect that hits the web at high speed. Spiders also can produce a glue-like substance to trap their prey in the web. Spiders also have the skill to avoid the glue when they traverse the web.

Engineers would love to harness silk for products such as bullet-proof vests, surgical thread, and artificial ligaments. A fortune awaits any team that can manage it. But no one has been able to do it. The materials engineering is simply too advanced.

Another remarkable biological material is the collagen found in animal tissue. Collagen is the main structural material in the different connective tissues within the body, including the bones, cartilage, tendons, ligaments, and skin. Its special triple-helix structure consists of amino-acid chains bonded together, creating a structure ideal for tailoring to different applications. This is accomplished by fine-tuning collagen's various parameters, including the degree of mineralization and how the collagen fibers are aligned. Collagen tissues can be rigid (bone), compliant (tendon), or possess a range of stiffness (cartilage). Some twenty-eight types of human collagen have been identified. Scientists are keen to develop materials based on the structure of collagen because of what Nan Zhang and his co-authors describe as the "exceptional properties of collagen: excellent mechanical strength and flexibility, biocompatibility, biodegradability, and ease of synthesis into a broad variety of multiscale hierarchical morphologies."[3]

Chitin is another notable material. Widely employed in the bodies of insects, it is ideal for their structures and hinges, and its rigidity and elasticity can be adjusted for various needs, fine-tuning carried out through the inclusion of specific minerals or proteins. Other impressive materials found in the biological realm include keratin (feathers), nacre (shells), and cellulose (wood). In each case the material is fine-tuned to produce the optimal properties for its application.

One reason the biological realm is able to contain so much ultimate engineering is that the materials used in each system are ultimate materials.

Ultimate Structures

Nature also contains breathtakingly optimal structural concepts. Figure 19.2 shows three important structural concepts found in biology. The last column shows equivalent structures used by engineers. I have published research papers on these structures.

HIERARCHICAL STRUCTURES

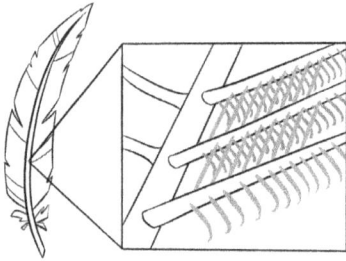

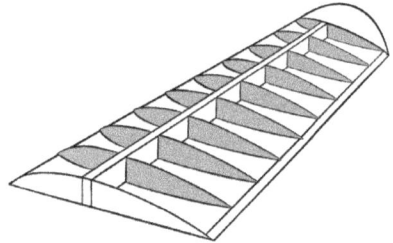

Feather   Aircraft wing

ULTIMATE BEAM SHAPE[4]

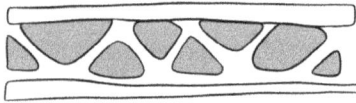

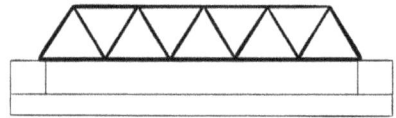

Bird bone   Warren truss

ULTIMATE COMPRESSION STRUCTURE[5]

Block-arched structure
formed by wrist bones
(shaded gray)

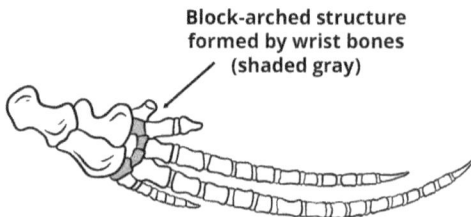

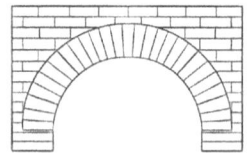

Whale wrist   Roman arch

Figure 19.2. Optimal structures in nature and manmade technology.

The first row, hierarchical structures, shows structures involving one or more main beams connected with an array of secondary, and sometimes tertiary, beams. Hierarchical structures are efficient because they bring loads from a wide area into a concentrated point with the minimum of material. They're commonly used in aircraft wings and bridges. Feathers are a particularly mass-efficient hierarchical structure, a key reason birds are so good at flying.

I have experimented with gull feathers in the lab and have been astonished at their lightness and robustness. Flight feathers have an optimal hierarchical structure in terms of load-bearing but also aerodynamics. The central shaft of feathers is called the rachis, with many barbs branching off each side. Very thin barbules then branch off from the barbs and are held together by small hooks that give the feather its airfoil shape. The rachis is hollow and tapers to form an optimal structure.

The second structure shown in the figure is the Warren truss. When I carried out research at Cambridge University, I wrote some research papers on optimized structures and showed why the Warren truss is optimized for applications where there is constrained height, as with bird wing bones.[6] Biologists were astonished when they found this optimal layout inside the wing bones of birds. For a bone to grow these diagonal struts requires very detailed and specific information in the DNA.

The third concept shown in Figure 19.2 is an arch structure with small compression elements. Since Roman times it has been known that the arch is an ideal structure for carrying loads when using materials that are strong in compression, such as stone or bone. The table shows how the whale has a block arch in the wrist bones for transferring loads from the digits to the lower arm.[7]

Ultimate Mechanical Linkage Mechanisms

Linkage mechanisms are used in many products, including vehicle steering units, vehicle suspension systems, crane-level luffing systems, angle-poise lamps, and double glazing window hinges. Figure 19.3 shows some key linkage mechanisms found in biology, with equivalent examples from engineering shown in the right column.

SHORT COUPLER AND EXTENDED BAR

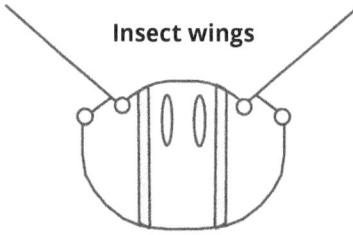

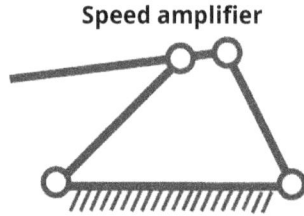

Insect wings   **Speed amplifier**

INVERTED FOUR-BAR MECHANISM

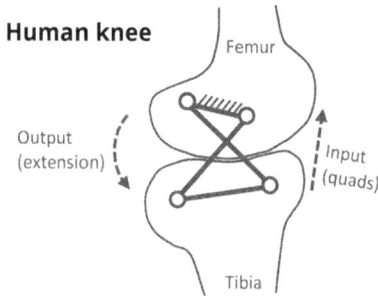

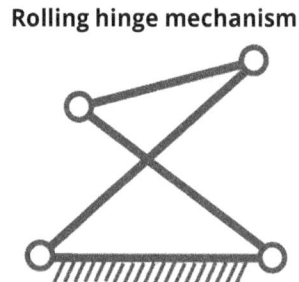

Human knee   **Rolling hinge mechanism**

Femur

Output
(extension)

Input
(quads)

Tibia

OVER-CENTER FOUR-BAR MECHANISM

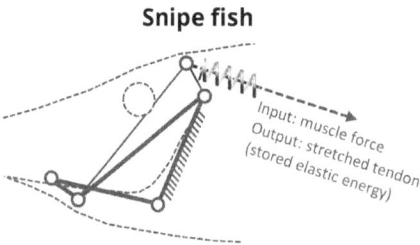

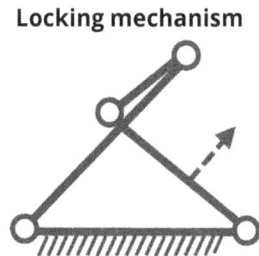

Snipe fish   **Locking mechanism**

Input: muscle force
Output: stretched tendon
(stored elastic energy)

REMOTE GROUNDING

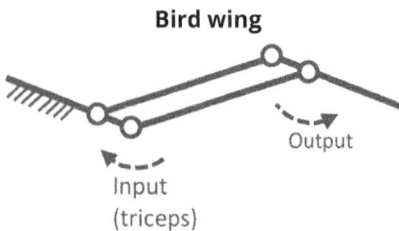

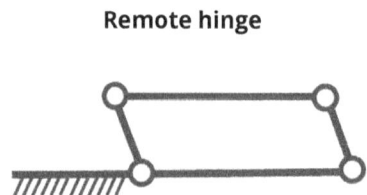

Bird wing   **Remote hinge**

Output

Input
(triceps)

Figure 19.3. Important linkage systems in nature and engineering.

Linkage mechanisms serve the crucial role in mechanical systems of optimizing forces and motions. In 2021 I wrote a review paper for the Institute of Physics (IOP) describing linkage mechanisms in biological systems.[8] The paper showed that biological systems use linkage mechanisms to create precisely optimized motions and forces. Figure 19.3 summarizes four of the main linkage mechanisms shown in that research paper. The ability of these linkage mechanisms to fine-tune mechanical performance comes compliments of the very large range of their possible layouts.

The first linkage mechanism in Figure 19.3 amplifies movement. By making the top (coupler) link relatively short, the link's rotation is greatly amplified for a given rotation of other bars. We find this feature in many insect hinge joints. The second example in Figure 19.3 shows an inverted four-bar arrangement (discussed in Chapter 2), which creates a compact hinge in both human and animal knees. This type of hinge, recall, affords a large range of motion in the knee joint. This joint is not the only one to use a linkage mechanism for optimal performance. As shown in Chapter 1, the lower leg has a linkage mechanism (parallelogram four-bar mechanism), and the lower arm has a linkage mechanism (a crossover four-bar mechanism), each one ideally suited for its particular role. Thus, human and animal joints have all the hallmarks of having been fine-tuned using advanced engineering.

The third row shows linkage mechanisms that can self-lock, as with locking wrenches. Incredibly, this clever human invention is found in snipe fish. The fourth row depicts remote hinges. When one bar is extended, it can be anchored behind a pivot point, allowing the mechanism to be conveniently supported. This support strategy is seen in bird wing mechanisms.

My IOP paper explains why the linkage mechanisms found in biological systems are more advanced than those built by human engineers and are, indeed, optimal. One reason is that manmade linkages are almost always two-dimensional, aligned in one plane. In contrast, biological linkages are often three-dimensional, with complex movements. For example, the knee joint undergoes very complex three-dimensional movements that enable the knee to lock when standing.

The fact that linkage mechanisms are seen at all in biology is evidence for intelligent design, because linkage mechanisms are irreducibly complex. And the fact that biological linkage mechanisms are superior to manmade engineering linkages makes the case for intelligent design all the stronger.

Ultimate Echolocation Technology

Human engineers would love to be able to match the echolocation technology found in creatures such as bats, whose sonar technology far exceeds any human-engineered version for resolution, dynamism, and adaptability. Echolocation represents another remarkably sophisticated type of engineering found in various animals.

Many creatures that are nocturnal, burrowing, or ocean-dwelling cannot use sight to locate food and instead rely on echolocation. These include bats, toothed whales, a few birds, and some small, burrowing mammals. The methods behind echolocation are various and include vibrating the throat and flapping the wings.

Echolocation occurs when an animal emits a sound wave that bounces off an object, returning an echo that provides information about the object's distance and size. Echoes can indicate the size, texture, and distance of the prey and other objects, such as surrounding walls. The cochlea, a large spiral cavity in the inner ear, is sensitive to a broad hearing range and finely tuned for the returning echoes.

Some bats can detect flies little more than one-hundredth of an inch across. Incredibly, bats can calculate the distance, size, and shape of objects like insects. Because insects are always on the move, bats have to click continuously, sometimes sending out almost two hundred calls a second. This and other echolocation systems in the biosphere appear to be at the limit of performance.[9]

When I was head of the department of mechanical engineering at Bristol University, my department had one of the world's leading research groups for ultrasonic measuring equipment. Despite the advanced systems we developed, we were keenly aware that our devices were slow, bulky, and inaccurate compared to those found in nature.

Ultimate Navigation Technology

Of all the wonders of biology, animal navigation is to me among the most amazing. Navigation is particularly important to flying creatures because they cover long distances and their view can be obstructed by mist, cloud, or dense forest. In addition, they can be blown off course by the wind.

The navigational ability of golden plovers has astounded scientists because of the length of flight and the danger of slight errors in direction. The birds' flight includes three nights of darkness and yet they follow an exact path towards Hawaii. Their navigation system appears to be pre-programmed. Using a magnetic compass based in their eyes,[10] the birds follow the magnetic field lines of the earth at a precise angle that leads to Hawaii. During the day the birds can also use the sun to correct their path. If the wind blows them off course, they adjust direction accordingly.

Homing pigeons employ a range of navigational cues, including the sun, Earth's magnetic field, scents, and landmarks. Honeybees, we now know, navigate by a combination of the sun, the polarization pattern in a relatively clear sky, and Earth's magnetic field.

Many marine creatures, including salmon, whales, and sea turtles, also perform incredible feats of navigation. Salmon, it is thought, navigate using Earth's magnetic field, and when they find the river they came from, employ memory and their keen sense of smell to navigate to their home stream. It is estimated that salmon can detect chemicals down to one part in a billion, an astonishing ability.

The Arctic tern has the farthest known migration of any bird. They fly from their Arctic breeding grounds to the Antarctic. In 2016 an Arctic tern was recorded clocking an incredible 60,000 miles over the course of its yearly migration.[11]

The monarch butterfly has one of the longest and most complex migrations of any insect, covering some 3,000 miles in an annual round trip from Mexico to the northeastern United States and southern Canada. On average it takes four generations of butterflies to complete the journey, which raises the tantalizing question: How does each offspring know what journey it is supposed to make? This question has not yet been fully answered, but part of the answer may be that the butterfly somehow knows to look for cues in each location, such as the type of food in that area.[12] This explanation, of course, opens on to deeper mysteries. How does a creature with a pinhead-size brain know this and pass it on to a descendant that the individual butterfly will never even meet, and pass on some information it will itself never employ or gain through experience, since its entire life will be spent on a different part of the migration path than its descendant's?

The globe skimmer dragonfly goes even further, from India to Africa, again using multiple generations to complete the journey.

Such extremes of migration[13] beggar belief on an evolutionary framework but fit readily within a paradigm in which, again, one anticipates the handiwork of a master designer eager to push the creative envelope.

Ultimate Molecular Machinery: The Flagellum

The bacterial flagellum is a highly sophisticated multicomponent nanomachine that performs the function of a screw-propellor system for many species of bacteria. This propellor system enables the bacteria to swim relatively long distances. The propellor is a helical flagellar filament and is driven by a proton-powered motor embedded in the cell wall and cell membrane. A counterclockwise movement drives the organism forward; a clockwise movement functions to change directions. The flagellum protrudes from the bacteria cell and is around 5–20 micrometers long. It can spin at tens of thousands of RPM.

The flagellum is made of many precision parts, some of them pictured in Figure 19.4. The long helical filament connects to the basal body via a flexible hook. This hook is the equivalent of a universal joint, commonly used in human engineering technology. The motor that drives flagellar rotation is powered by the flow of protons across the cytoplasmic membrane. If all this weren't sophisticated enough, there is also the matter of automated flagellar assembly, a process of outlandish sophistication occurring in several stages and involving many genes.

When engineers study the flagellum, inevitably they come away in awe of the precision and efficiency of the machine. It is an excellent example of irreducible complexity and ultimate engineering.

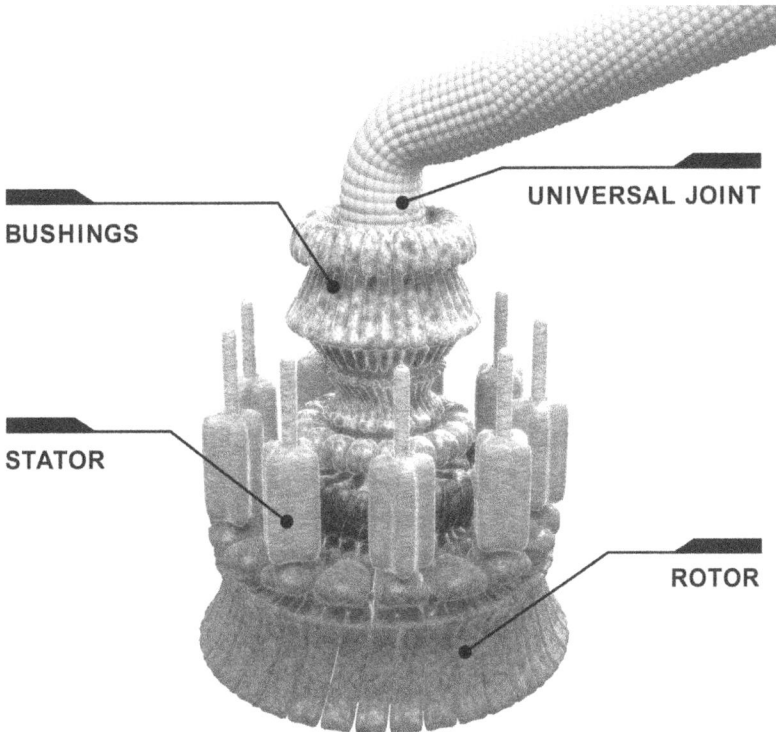

Figure 19.4. The bacterial flagellum.

Ultimate Lubrication

I have worked for the British Olympic cycling team for three Olympics—Rio, Tokyo, and Paris. For each Olympics we tried to produce the best lubrication system in the world to create the lowest possible friction. We believe we succeeded, though to be precise, I should say that we created the best lubrication system in the Olympics. It was not the best lubrication system in the world, because it was not as good as the synovial lubrication system found in animal joints.

After the bicycles with our gear system topped the cycling medal table at the Rio Olympics, I decided to look up the friction levels in synovial joints for comparison. I long ago had become habituated to uncovering examples of ultimate design in nature, and yet when I read the data for human joints, I was stunned at what I read. The synovial fluid's (kinetic) friction coefficient has been measured to be as low as 0.002,[14] meaning the friction force is 1/500th the normal force. This friction level is fifty times better than that of our world-leading lubrication system. This is astounding and helps explain why a healthy animal joint experiences virtually no wear after millions of cycles.

When joints experience short-term impact loads, during activities like running, the synovial fluid is trapped between the joint surfaces, leading to a squeeze-film effect that shields the surfaces from direct contact and acts as a shock absorber. In cases where the squeeze-film effect is insufficient for the impact, the surfaces are protected at the molecular level by the synovial fluid's large protein molecules. Cartilage also plays a role in the lubrication system in the way it acts like a sponge in absorbing synovial fluid and releasing it when the joint becomes loaded. At present engineers cannot copy the synovial fluid system because of the complexity of the materials and molecular processes. Synovial fluid is, unquestionably, ultimate engineering.

Ultimate Engineering Points to Intelligent Design

Some evolutionists claim that a blind evolutionary process can produce extremes of design if given enough time. However, such faith in time is not justified, not only because the math does not support the argument,[15] but more obviously, because as Nathan Lents, Abby Hafer, and Richard Dawkins have rightly emphasized, where the evolutionary mechanism of random mutation and natural selection does manage to create something new, the mechanism can be expected to produce a substandard design, even when a vast amount of time is available. So the mechanism is a poor fit with occurrences of ultimate engineering in nature.

Atheist biologist Steve Jones commented, "Natural Selection is a machine that makes almost impossible things" and that new designs appear "as if by magic."[16] Jones is correct to describe nature as containing designs that are almost impossible. However, to say that natural selection is a machine that can do impossible things is nonsensical. As explained in the introduction, engineers know very well that selection by itself can create precisely nothing. Only when selection is accompanied by a means of generating concepts can selection achieve anything. While the process of natural selection can facilitate microevolution, it cannot create novel designs.

If you read books on evolution by its public defenders, notice how they never give detailed scientific explanations for the origin of subsystems like the blood circulation system or the linkage mechanism in the knee. Their explanations are often little better than "natural selection did it." To say "natural selection did it" smacks of god-of-the-gaps logic. It has no scientific content. In this case, the god is Dawkins's blind watchmaker, the evolutionary mechanism. Such statements are statements of faith in magic; they are not science.

When I invented the concept of the double-action worm gear set for the Envisat and MetOp satellites, the novel design was the result of intense creative thought. There were no ready-made sets of concepts for me to select from. If there had been, my job would have been incredibly easy. But that isn't how invention works. Exactly the same principle applies to biology. Natural selection cannot drive macroevolution because evolution's means of generating novelty, random variation, is wholly inadequate to the task of searching a practically infinite sea of possibilities for those rare and intricate configurations that achieve genuinely novel form and function.

When you study nature, you do not find an insect species with a suboptimal form of flight. You do not find a simple echolocation system. Instead, apart from the effects of aging and genetic decay, you find optimization and ultimate engineering sophistication beyond anything human engineers have achieved. Evolutionary theory fails to identify a cause with the demonstrated ability to produce these radically novel and ingenious design concepts that have popped up innumerable times in the history of life. The scientific theory of intelligent design readily accommodates the discovery of ultimate engineering.

And when it is yoked to a theistic framework—what I have referred to as *theistic design* for simplicity—then ultimate engineering in biology is not merely well accommodated but is positively predicted.

The Explanatory Power of the Intelligent Design Framework

The theory of intelligent design, at its core, is a theory of design detection coupled with the observation that certain features of the natural world carry the clear signature of having been designed for a purpose. The theory makes no demands on how well or poorly the thing in question was designed. After all, one could use the tools of design detection (not to mention those of simple common sense) to ascertain that a shoddily constructed bicycle, discovered in the woods, was the work of a designing intelligence rather than the product of mindless natural forces. Poor design does not equal no design. At the same time, the theory of intelligent design is open to the possibility that the various biological forms were designed with ultimate engineering. Openness to this possibility is reinforced by the repeated discovery of masterful designs in biology, leading design theorists to positively anticipate the discovery of more examples of ultimate engineering in the living world. (As noted earlier, a specifically theistic design framework further encourages this anticipation of ultimate engineering.)

A design perspective also anticipates that all organisms are subject to decay of genetic information over successive generations. Therefore, there is the expectation of finding genetic decay in the human species. The speed of decay is slowed by the moderating action of natural selection and sophisticated repair mechanisms at the cellular level, but they do not completely halt the decay. To recognize ultimate engineering today, it is necessary to identify and take account of the effects of this genetic decay. At the same time and despite genetic decay, proponents of intelligent design generally anticipate that the human body still has a level of design sophistication far above human technology.

In contrast, evolutionary theory holds that the line leading to modern humans has gradually gained in sophistication from relatively primitive ancestors. Evolutionary theory also predicts that the human body will be found to be marked by substandard design, inferior to human technology, due to the severe constraints on the evolutionary process. Evolutionary theory does allow for genetic decay where environmental factors have reduced the effectiveness of natural selection,[17] and it allows for particular biological forms to get stranded on fitness plateaus that make further upward evolution difficult or impossible; but the larger pattern commonly emphasized by its adherents is one in which an organismal form tends to evolve from the less sophisticated to the more sophisticated over time, even as the evolutionary mechanism produces substandard designs along the way.

Figure 19.5 shows a simplified schematic graph of what intelligent design reasoning and evolutionary theory predict over time for the level of design sophistication in the human body. (The graph could be given for other organisms.) As shown in the first part of this book, scientific evidence confirms the prediction of ultimate engineering in the human body.

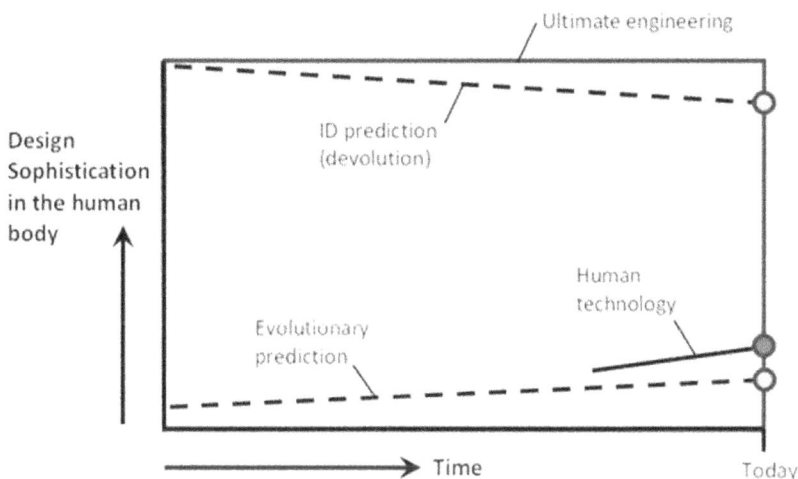

Figure 19.5. Predictions of how the human body (and any purported hominid precursors in its line) changes over time.

We often hear claims that a design framework cannot be used as an effective explanatory tool. But it is just such a framework that is able to explain, and even comes to anticipate, the reality of ultimate engineering in biology, whereas modern evolutionary theory struggles to come to terms with the repeated discovery of extreme levels of design sophistication in biology.

In the next chapter we will see how the design framework also better explains the extreme diversity of designs we find in nature.

20. Ultimate Diversity Throughout Nature

A Theistic Design Paradigm Anticipates Ultimate Diversity

The previous chapter argued that a theistic design paradigm correctly anticipates ultimate engineering in biology. Here I want to argue that this designing intelligence can also be expected to produce extremes of diversity in the biosphere, including the following:

- Extreme diversity of sizes.
- Extreme diversity of locomotion types.
- Extreme diversity of color-producing mechanisms.
- Extreme diversity of habitat.

This is an inference drawn partially from observing human designers. I know from my forty years of experience in engineering that designers like to explore extremes of design. For my own part, I found it very fulfilling working with the British Olympic Cycling Team to design the fastest track bike in the world. I still have bicycle components on the display shelf in my study that were used to break various world records. It was also satisfying to help design the world's largest civilian spacecraft, Envisat. I also keep components from it on my study display shelf.

Approaching the subject theologically, from within the Judeo-Christian framework that served as the soil for the scientific revolution, one understands that humans are made in the image of God, a truth we are told in Genesis immediately after learning of God's mighty acts of creativity. We are creators made in the image of *the* Creator, so if we delight in pushing the creative envelope, it's quite reasonable to anticipate that he does as well, even if at a much higher level.

It makes perfect sense then to suppose that God, having performed the Herculean task of creating a planet for life, would seek to explore extremes of design for pleasure and to demonstrate his wisdom and power. Another motivation for designing extreme diversity is to provide an abundance of organisms for humankind to use and appreciate. Yet another reason for diversity is to give examples of brilliant design for humankind to copy.

In sum, the theistic design perspective anticipates both ultimate engineering and extreme diversity throughout nature. As we have seen and will see, this is precisely what we find on planet Earth. The founders of modern science undertook their investigations of nature with very much this expectation, which led them to anticipate hidden and ingenious designs in nature, and to go looking for them.

In contrast, extremes of design are not what the evolutionary paradigm anticipates. This is because the evolutionary mechanism is a mindless process that has no motivation to explore the limits of design. Extremes of design require highly advanced engineering solutions, and the theory of evolution does a poor job of explaining why any such should appear when evolutionary processes would tend to evolve only what is "good enough" for survival and reproduction. The mantra that "evolution fills every niche" is not based on science but on a dogmatic assertion that "evolution must have done it because it is here."

This chapter gives a few examples of extremes of diversity in the natural world and how these fit both the minimalist theory of intelligent design and a more expansive theistic design paradigm.

Exploring Extremes of Size

There is scientific evidence that the largest creatures that have lived on earth (in land, sea, and air) are at the limit of what is physically possible.

Dinosaurs are the largest animals to walk on land. Based on fossil evidence, it is estimated that the largest weighed around seventy tons and were over forty meters long.[1] That is more than five times the weight of the largest African bush elephants. There are biomechanical reasons why the largest dinosaurs are thought to be at the limit of what is possible.

When researching at Cambridge University, I wrote a paper on scaling effects on organisms that placed limits on size.[2] For example, if an object is scaled up in length by a factor two, then the volume and weight go up by a factor of eight (2^3, because volume has three dimensions). However, the cross-sectional area of parts like legs only goes up by a factor of four (2^2, because area has only two linear dimensions). This means that as objects scale up, either the components must be relatively large, or the materials must withstand higher stresses. This is why there is a limit to how big bridges can be.

Exactly the same principle applies to large animals like dinosaurs. As creatures scale up in size, the weight scales up at a faster rate than the cross-sectional area of limbs. As a creature scales up in size, so the limbs become more highly loaded and need to become relatively large in cross-section. This explains why elephants have very broad legs and short feet for their size, and why dinosaurs had even broader legs relative to their size. Because of scaling principles, there is a hard limit to how big a land creature can get. And in the case of the largest dinosaurs, it is thought that they are at that limit.

To reach the limits of size, large dinosaurs had some extremely sophisticated design solutions. For example, they had pneumatic air sacs in their bones. One advantage of these is that it makes skeletons lighter. A second advantage is that it produces a more efficient one-way air breathing system similar to that of birds. This one-way air flow involves pathways connecting the air sacs with the lungs.

Because of the principle of irreducible complexity it is very hard to explain how the complex air sac system in dinosaurs could have evolved by chance and natural selection. To produce such a system would involve many changes from the two-way breathing system of smaller land creatures, changes that would have to originate together for the system to function. In contrast, the extremes of size and the special systems that make it are well within reach of the intelligent design framework, since a conscious, purposive designer can look ahead to a future goal and set about fashioning and assembling various parts with the functional goal in mind, without having to creep through countless generations of an animal while adding one little addition at a time.

In the case of the marine environment, we find even larger creatures than dinosaurs because of the way water provides support to the body. Blue whales are the largest known creatures to have lived on earth, weighing up to two hundred tons with lengths up to around a hundred feet. Studies indicate that blue whales are at the limit of size because of demands placed on the heart to pump enough blood to all the tissues of such a large body.[3]

Flying creatures from ancient times also appear to have been at the maximum size that is physically possible. The largest flying creature to have ever lived is thought to be *Quetzalcoatlus*, a member of the ancient group of flying reptiles called pterosaurs. This reptile had thin limbs, a long beak, and a gigantic forty-foot wingspan.

As with land creatures, there is a scaling effect that limits the maximum size of a flying creature. As a flying creature is scaled up, the weight goes up with the cube of the length, whilst the wing area only goes up with the square of the length. And it is the area of the wing that has to support the weight of the creature in flight.

This is why larger birds have relatively larger wing areas. And this is why there is an upper limit on the size of a flying creature. There comes a point at which it is not possible to support the body weight by the wings. Also, as birds get larger, the speed required for take-off gets larger and so this requirement is another constraint on size.

The fact that there are or have been land, sea, and air creatures as big as is physically possible is a problem for evolutionary theory, which as discussed, is restricted to blind gradualistic searches and a good-enough-is-good-enough ethos. But it is well accommodated by intelligent design and, again, positively anticipated by a theistic design framework.

In nature we also see creatures at the limit of what is possible for miniaturization. One of the smallest land creatures is a beetle, *Scydosella musawasensis*. It was first discovered in Nicaragua and is around 0.3 millimeters long, which is about twice the thickness of a strand of coarse human hair. The smallest multicellular marine creatures are thought to be the myxozoan invertebrates, which when fully grown can be as small as 8.5 μm (less than a hundredth of a mm). Among the smallest flying creatures are fairyflies, which are tiny wasplike insects. They range mostly from 0.5 to 1.0 mm long, but some have a body length of only 0.139 mm.

Living at these extremely small scales creates major engineering challenges. Locomotion is difficult because the laws of fluid dynamics mean that moving through air can feel like moving through treacle to a tiny organism. Another challenge is that of constructing miniature components such as legs, wings, eyes, and mouths. Such parts require hinges and sensors on a microscopic scale. Because of these engineering constraints, it is thought that these creatures are at the limit of miniaturization. This, too, is well accommodated by the theory of intelligent design and positively anticipated by a theistic design paradigm.

Exploring Extremes of Locomotion Types

Ground locomotion comes in many forms, including walking, running, hopping, crawling, swinging, and jumping. For each type of locomotion, there is an extraordinary variety of designs. The range in number of legs is astonishing. At one extreme is bipedal walking, as with humans. At the other extreme is the millipede *Eumillipes persephone*, which has 1,306 legs.

An extreme variation in walking is found in mountain goats, which can walk on tiny ledges on near-vertical cliff faces. Running sometimes contains quite a variety of forms in one creature. Horses usually have four gears (gaits)—walking, trotting, cantering, and galloping. These gaits aren't just different speeds, but involve distinct footfall sequences and biomechanics. Snakes have at least four types of crawling, including sidewinding, concertina, serpentine, and rectilinear. In the case of apes and monkeys, their main form of loco-motion involves knucklewalking, with many species possessing the additional ability to travel adroitly through trees using their forelimbs to swing and jump. The planthopper insect (*Issus coleoptratus*) uses a gear-driven mechanism to hop great distances.

We find an astounding variety of flight mechanisms in the living world. The engineering skill evident in these is without parallel in the realm of human design. This fact is easily overlooked if one simply focuses on, for example, the speed of the fastest bird versus the speed of the fastest airplane. But engineers readily acknowledge that the engineering behind flight in nature greatly surpasses what human inventors have managed. Animal flight is far more stable and able to tolerate disturbances such as gusts of wind. Animal flight is also far more efficient, quiet, and adroit. And when you compare size for size, natural flyers always outperform human technology.

Birds generally use the lift of an airfoil for gliding and flapping flight. Among birds we see different species excelling in a great range of flight types. The albatross has the ultimate long-distance gliding design. Peregrine falcons are designed for great speed. Eagles and condors exhibit extraordinary strength, allowing them to lift relatively heavy prey into the air. Hummingbirds are the ultimate design for hovering.

The avian world also displays a great range of flying techniques. To save energy, some birds rise on a thermal and glide to the next thermal for the next free ride. Other birds perform burst-glide flight where they put in a burst of flight and then glide. Still others perform bounding flight, involving bursts of flight followed by parabolic coasting with the wings folded against the body.

Insects also exhibit various sophisticated flight styles, including airfoil lift, leading-edge vortices, and clap and flap. These techniques enable very small creatures not only to fly, but to fly with an agility that engineers cannot yet match.

We also find in the biological realm all manner of extreme designs related to swimming. The hydrodynamic requirements of swimming in water mean there is always a trade-off between speed and maneuverability. Depending on the environment, aquatic creatures range from slow and very maneuverable to fast with limited maneuverability. One of the slower swimming modes, used by fish such as eels, is the anguilliform, in which a long slender body makes a wave motion. However, such fish have incredible maneuverability and can navigate through small rock openings.

One of the fastest swimming techniques, used by fish such as tuna, is called the thunniform, where the body is mainly still but a large powerful crescent-shaped tail rapidly oscillates to produce high speed. Engineers have been astonished at the design and speed of these fish. Tuna have been recorded swimming at speeds up to 45 mph, similar to the top speed of white-tailed deer running over flat ground. Considering that the density of water is around a thousand times greater than air, it is incredible that a fish can swim as fast as a deer can run.

Marine animals use a remarkably wide range of swimming techniques. Some fish can swim using their pectoral or dorsal fins. There are also marine creatures, such as manta rays and penguins, that flap their fins (or flippers) under water, much as birds flap their wings in flight. Cephalopods like squid commonly use jet propulsion by filling their mantle cavity with water and then quickly expelling the water to produce a jet force.

Recent research has revealed advanced design features that enable fish to swim with extreme speed and efficiency. For instance, it's been discovered that fish fine-tune stiffness during each swimming stroke, enabling them to store and recoup energy during fin movements.[4]

Also impressive, eels and mackerel become elastic in different parts of their bodies and at different points during each undulating cycle. Engineers would like to copy this special feature but are finding it extremely difficult to emulate.

Some creatures are like decathletes in the way they can perform multiple forms of locomotion. In the case of some water birds, they can fly in air, walk on ground, and swim underwater by flapping their wings. In 2014, I co-wrote a review paper in the journal *Bioinspiration & Biomimetics* on multi-model locomotion in nature, showing that many animals are incredibly well designed for multiple forms of locomotion.[5]

The great range of locomotion techniques and their advanced design features are a beautiful fit with the theory of intelligent design, a paradigm that allows for the work of a highly creative master inventor in the history of life.

Exploring Extremes of Color-Producing Mechanisms

Many animals produce color through pigments, such as the melanin pigment that produces browns and blacks in human skin and the carotenoid pigment that produces pink in flamingos. However, many animals also feature incredibly advanced engineering solutions to produce structural colors. Structural coloration can be created by microscopically structured surfaces, thin-film surfaces, or small particles within transparent materials. In some cases, structural coloration occurs in combination with pigments. Each of these structural coloring methods produces high-quality colors and patterns, and each involves astonishingly precise engineering.

- **Photonic nanostructures,** found in many beetles, birds, and butterflies, involve precisely spaced surface features, spacing comparable to the wavelengths of visible light (400 to 700 nm), which cause a variation in the refractive index. When white light reflects off the surface, color is produced due to the variation in the refractive index. Because the scale of the structures is so small, different colors and patterns can be produced in fine detail, as with butterfly patterns. Engineers have been astonished at the precision of the nanostructures in creatures such as butterflies. It is hard for the evolutionist to explain why butterflies need to have such exquisite patterns, or could have them, especially when such exquisite engineering is needed to produce them. But the fact is readily accommodated under the design paradigm.

- **Thin-film interference** is produced when light waves are reflected by the upper and lower boundaries of a thin film.

When the thickness is comparable to the wavelengths of colored light, certain wavelengths are removed or added, and this changes the color of the reflected light. An iridescent effect (changing color) results when the angle of view is changed. The buttercup flower has an unusual thin-film effect. The glossy appearance of the petals is due to a yellow-pigmented epidermal layer that acts as an optical thin film.[6] Because the epidermal layer is very smooth and thin, optical interference occurs and the reflected light creates a glossy sheen. The buttercup is considered a simple flower, and yet its beautiful color stems from meticulously precise engineering.

- **Light scattering** is the process by which small particles scatter light, causing transparent materials to exhibit colors, especially blue. The color of blue eyes is due to the scattering of light by a translucent layer in the iris that contains numerous particles less than 1 micron in diameter. This scattering is more pronounced for shorter wavelengths, such as blue and violet, than for longer wavelengths, such as red and orange. As a result, when there is no pigment, eyes appear blue. When there is a small amount of melanin pigment, hazel and green can be produced. The structural blues in bird feathers are often created by the scattering of light from the air bubbles within the spongy matrix of the barbs' keratin.

- **Bioluminescence** involves a process by which a creature makes light from a complex chemical process. Bioluminescence is found in a wide range of animals, including insects, centipedes, millipedes, snails, earthworms, echinoderms, fungi, fish, and squid, as well as some microbes. One of the most widely seen examples of bioluminescence is in the firefly. The chemical reaction that turns energy into firefly light is a marvel of efficiency,[7] far better than that of incandescent bulbs (around 5–10 percent). Bioluminescence requires three things: (1) a light-emitting pigment, known as a luciferin; (2) a complex enzyme protein called luciferase to cause the reaction; and (3) precise neurological control. Explaining bioluminescence by evolution is extremely hard because many complex elements have to be in place at once. Yet bioluminescence is supposed to have arisen via gradual evolutionary processes at least thirty times in evolutionary history.[8]

- **Metachrosis** is arguably the most astounding type of coloring. This smart color-changing mechanism enables animals like squid to rapidly adjust their color for camouflage in different settings. Some creatures can even mirror back the colors of the environment—almost as good as an invisibility cloak. Animals that can change color include octopuses, cuttlefish, and squid (cephalopods). If you want to be dazzled, jump onto YouTube and search for "octopus camouflage" or "squid camouflage." Metachrosis requires thousands of color-changing cells, such as *chromatophores*, just below the surface of the skin. The complex color-changing mechanisms include reversible proteins that can be switched between two configurations. This can be controlled by electric charge. Sometimes muscle contraction is used to expose certain colors. Since chromatophores contain multiple reflecting layers, the switch changes the layer spacing and hence the color of light reflected. Cephalopods can change not only their coloring but also the texture of their skin to match nearby features such as rocks and corals. They do this by controlling the size of projections (called papillae) on their skin, creating textures ranging from small bumps to tall spikes. Human engineers are nowhere near being able to create such high-quality camouflage that is simultaneously this adaptable. Once again we have a case of ultimate engineering that evolutionary theory struggles to explain but that is well explained with the idea that such creatures are the work of a master designer.

- **Lasers** emit beams of monochromatic light… in nature? As part of the design of an animal? Yes. This last discovery is so extraordinary it reads like a spoof, but it's not. The laser was invented in 1960, with the inventor, Theodore Maiman, building on the foundational work of Nobel Laureates Albert Einstein, Charles Townes, and other geniuses. But it turns out that Maiman was merely reinventing the laser. As in so many other cases of human technological innovation, that most intelligent of designers got there long before. In 2025, researchers made the incredible discovery that peacock feathers have structures that produce laser beams. They don't generate the light themselves, but the eyespots have unique properties that align light waves by bouncing them back and forth, effectively turning them into yellow-green lasers.[9] When engineers produce laser beams, they use complex equipment, and yet a peacock feather can perform this complex process by itself.

Evolutionary theory does a poor job of explaining why animals feature such a myriad of advanced design solutions for producing high-quality colors. But the finding is beautifully in keeping with a theistic design perspective, which anticipates that the master artist behind the world would pursue excellence in this way.

Exploring Extremes of Habitat

One of the most impressive examples of remote living involves life in the deep sea. In 2017 the BBC produced a TV documentary called *Blue Planet II: The Deep*. The program explained that scientists had not expected to find life forms in the deep, because of the total blackness and extreme pressures there. At depths of 4,000 meters the pressure is around 2.67 tons per square inch. (Imagine a creature that is a square inch across feeling a pressure equivalent to 2.67 tons resting on it.) But find them they did.

Any creature at that depth requires extreme engineering solutions to survive. The submarines that filmed the deep-sea creatures were made of solid thick titanium tubes with incredible strength. And yet they were filming living creatures that thrive and reproduce in the deep. It is not surprising that the scientists did not expect to find so much life in the deep oceans, because it appears that most or all of them involved in the work embraced what they were taught in school, evolutionary theory, which anticipates poor design in nature or, at best, designs that are just good enough. A theistic design framework, in contrast, anticipates a cosmic designer interested in pushing the envelope, which is just what we find in the case of these extreme creatures.

The emperor penguin that thrives in Antarctica is another example of a creature living in an extreme location. It requires very sophisticated design features to live through a four-month winter at -50°C with no access to food. Another animal that is brilliantly designed for the cold is the Myanmar snub-nosed monkey, discovered in 2010. It has woolly fur, prominent red lips, and an upward-facing nose. It lives in temperate cloud forests in the eastern Himalayas, at altitudes ranging from 5,700–10,500 feet and where temperatures are often freezing.

One of my favorite creatures is the markhor goat. It also lives in the Himalayas, and it has an incredible ability to climb near vertical cliff faces due to an extremely fine sense of balance and unusually grippy feet. Another favorite creature designed for extremes is the camel, with its ability to survive long periods in drought conditions.

An amazing example of remote living is found with the seals that live in the remote Lake Baikal in central Asia. When pups first move out of their birthplace, they must swim with their mother a long distance under the ice. The mother has no problem managing it, but it is often too far for the pup to manage in one breath. Recent observations have shown that the mother seal will stop in the water and blow air bubbles to form an air hole under the ice for the pup to take a breath along the way.

This array of examples of creatures living in extreme habitats is again exactly what would be expected from theistic design, where a master designer seeks to explore a dazzling range of design solutions.

Some bacteria live in such hostile environments that the bacteria have been dubbed extremophiles, the very name underscoring that they are designed at the limit of what is possible. Extremophiles are capable of living at extremes of temperatures, pressures, radiation levels, salinity levels, or pH levels. Some of these microbes can even survive more than one type of extreme, earning them the name polyextremophiles. Extremophiles and polyextremophiles have surprised modern science because they have been found "where nobody expected life to survive, let alone thrive."[10]

Researchers at Ghent University in Belgium have presented findings that show a species of Bacillus bacteria survived after being heated to temperatures of 420°C.[11] Some bacteria also have been found living in the cold and dark in a lake buried a half-mile deep under the ice in Antarctica. Other bacteria, found in the Mariana Trench, the deepest place in the Earth's oceans, are able to withstand immense water pressures.

Theistic Design Anticipates Earth's Extremes of Diversity

Whereas evolutionary theory must scramble to throw up strained and ad hoc explanations for the countless manifestations of extreme diversity in the biosphere, such findings are readily accommodated by the theory of intelligent design, and more than this, strongly anticipated under a hypothesis of theistic design. From dinosaurs at the limit of how big a land animal can be to laser beams in peacock feathers, and the wondrous riot of extreme diversity in between, life points not to a mechanism incapable of thought, foresight, planning, and ingenious leaps of creativity but to a master designer interested in exploring variety in color mechanisms and locomotion, extremes of habitat and the limits of size great and small. Such a panoply of ultimate engineering points to a singularly intelligent design.

As a concluding aside to this chapter, I can't resist adding the observation that an intelligent designer such as we see coming into focus from the above observations could also be expected to explore a great diversity of designs of stars and galaxies. Therefore I would argue that the vast universe is also very much what a theistic design paradigm anticipates.

21. Will the Real Pro-Science Paradigm Please Stand Up?

Which Paradigm Better Follows the Scientific Evidence?

We are often told that evolution, not intelligent design, follows the scientific evidence. Having published more than two hundred scientific papers on the science of design in engineering and biology over the last forty years, I have concluded that it is actually intelligent design and not evolutionary theory which better follows the scientific evidence. Some senior biologists have openly admitted that evolutionary theory is a dogma adhered to regardless of what the evidence suggests. In Chapter 17 we saw Sir Julian Huxley, Harvard's Richard Lewontin, and Scott Todd (the latter in the journal *Nature*) all confessing to this, each in his own way. Unfortunately, students are taught this same inflexible attitude, that they can and should be confident in evolution despite the lack of evidence.

Over the last thirty years I have asked many biologists how they can be confident in evolutionary theory. One senior biologist (who was head of the school of biological sciences at a major university) replied, tongue-in-cheek, "Biologists are confident because it must have happened once, otherwise we would not be here!" I told him that this reasoning is not science but blind faith. He agreed and added that it is no better than a god-of-the-gaps belief.

Other biologists have said that they believe in evolution because most other biologists believe in evolution. Again, this is not a scientific explanation. Interestingly, some biologists have said that, while they can see that macroevolution appears impossible in their own area of specialization, they believe in evolution because apparently it works in so many other subdisciplines. When I point out that scientists in other fields, including other subdisciplines of biology, say exactly the same thing, they try to change the subject.

Some biologists tell me they are confident in evolution because natural selection is such a powerful force. But as noted in the introduction, selection by itself cannot create anything, because it can only choose among what has already been created, and therefore faith in natural selection is blind faith. An example of this attitude can be found in Richard Dawkins's *The Greatest Show on Earth*, where he says, "We have no evidence about what the first step in making life was, but we do know the kind of step it must have been. It must have been whatever it took to get natural selection started.... by some process as yet unknown."[1]

When students are told that a designer-free origin of the first life is a superior scientific explanation, it is simply not true. I have often challenged my academic colleagues on this point. When I say to them that students are being misled, they often agree. When I ask them how they justify students being misled, they tell me that it would be too complicated to explain that there is little if any evidence for abiogenesis and that it is far easier to just tell them it is a "scientific" theory so that the class doesn't get bogged down.

Such private confessions are corroborated by what physicist and astrobiologist Paul Davies stated publicly, namely that evolutionists "feel uneasy about stating in public that the origin of life is a mystery."[2] Remember, as well, biologist Steven Vogel admitting that evolutionists usually keep quiet about weaknesses in evolutionary theory: "We biologists recognize these [evolutionary] constraints, but we don't often rise above our natural chauvinism and make enough noise about them."[3]

There is a great need today for the scientific community to be honest with the public and with students about the lack of evidence for evolution. Charles Darwin himself often identified weaknesses in his theory, but today there is an unwritten rule that evolution cannot be questioned. Without transparency, students and the public are being seriously misled.

Engineering, Not Evolutionary Dogma, Is Key to Understanding Biology

Modern science claims that biology has two foundational subjects: chemistry and evolution. Theodosius Dobzhansky famously claimed that "Nothing in Biology Makes Sense Except in the Light of Evolution."[4] However, when it comes to abiogenesis and macroevolution, evolutionary theory offers anything but light. The phrase "evolution did it" is a mere dogmatic statement of the faithful, not a scientific explanation.

In reality, the two foundational subjects to biology are chemistry and engineering. Biology is full of advanced engineering, and it is thanks to the discipline of engineering that there have been so many recent advances in our understanding of biology. Instead of taking courses in evolutionary theory, biology students would benefit far more from engineering courses. It is interesting to note that abiogenesis is sometimes referred to as chemical evolution. In fact, chemical evolution, strictly speaking, refers only to the processes that formed the organic building blocks of life from simpler precursors. But the fact that the two terms, abiogenesis and chemical evolution, are routinely conflated is telling. The impression given is that life could readily appear just through a series of chemical reactions. Of course, chemistry plays a major role in life, but engineering plays an even greater role, because it coordinates everything from the molecular level to the macro level. The origin of life is not an act of chemistry; it is a monumental act of engineering design that employs various chemicals and chemical reactions.

To understand animal biomechanics, you have to understand engineering mechanics such as linkage mechanisms, energy storage, power amplification, and mechanical advantage. I once debated the origin of linkage mechanisms with a biologist. I asked about the inverted parallelogram four-bar linkage of the knee joint. He confessed he had no idea what a four-bar linkage was, let alone an inverted parallelogram four-bar linkage. I then asked, given that he had no idea about the biomechanics of the knee, how could he be confident it had emerged through blind natural processes?

Nathan Lents's book *Human Errors* claims the human body has hundreds of design flaws. But as shown in earlier chapters, his book is seriously lacking in knowledge about engineering principles relevant to biomechanics. It takes decades of study and practice for an engineer to become competent to review and judge an advanced design. And yet Lents attempts a design review of the human body without reference to any engineering design principles.

The Evolution Paradigm Is Bad for Science

In the first part of this book, I showed that when it comes to assessing design in the human body, evolutionists like Nathan Lents, Abby Hafer, Jerry Coyne, and Richard Dawkins repeatedly make the mistake of following an evolutionary worldview that predicts bad design, rather than following the scientific evidence that demonstrates ultimate engineering. When scientists inflexibly champion a dogma rather than following the evidence, scientific progress inevitably suffers. This is what we see today because of the evolutionary paradigm.

Of course, research on biological adaptation has been fruitful, with many excellent research studies, including investigations carried out by Charles Darwin himself in the nineteenth century. The problem is that Darwin mistakenly concluded that his work on micro-evolutionary adaptation could be extrapolated to macroevolution. That error remains with us today, and the framework leads his followers to expect substandard design in biology. The result is that when some as-yet poorly understood biological organ or system is considered, many evolutionists automatically conclude it must be badly designed rather than withholding judgment and studiously investigating form and function in order to make scientific progress.

In my own research on biomechanics, I have seen firsthand how evolutionary assumptions lead to false conclusions. Evolutionary theory is especially primed to see bad design in human joints be-cause, according to the paradigm, our bipedal style of walking evolved without planning from knuckle-walking quadrupeds. The first four chapters of this book have given many examples of false scientific claims by evolutionists about human joints.

My published research on the vertebrate limb has also exposed false evolutionary teaching. As explained in Chapter 17, it is common to read on museum websites the false statement that the whale flipper is a rigid paddle. These museums are perpetuating the false homol-ogy arguments of Charles Darwin that have long appeared in biology textbooks. This is a prime example of how evolutionary thinking holds back scientific progress. I look forward to the day when textbooks and museums correct this false teaching.

Another example of the debilitating effects of evolutionary think-ing can be seen with so-called "junk DNA," those non-coding regions of DNA that evolutionists wrote off as junk but that turned out to have vital functions, just as intelligent design proponents predicted. Evolu-tionists also have been quick to claim that the human eye is wired in backwards, degrading the quality of the light as it passes through the back of the retina. But as we saw in Chapter 9, scientific discoveries have shown that Müller cells act as fiber optic cables to transport light through the retina without signal degradation, and the "backward wiring" improves oxygen flow to the oxygen-hungry cells. These dis-coveries, I submit, could have been made sooner and more swiftly built upon by other investigators if there had been no assumption that the eye was poorly designed by a blind evolutionary process. Even today, the talking point about bumbling evolution's "backward wiring" of the vertebrate eye is repeated by various evolutionists, despite the claim having been thoroughly debunked. This illustrates how the damage done by evolutionary thinking can take a long time to clear up.

One might assume that, even acknowledging the errors evolution-ary theory has encouraged, surely it has more than made up for these with its far-reaching contributions to biology. National Academy of Sciences member Philip Skell investigated and came to a very differ-ent conclusion:

314 / Ultimate Engineering

My own research with antibiotics during World War II received no guidance from insights provided by Darwinian evolution. Nor did Alexander Fleming's discovery of bacterial inhibition by penicillin. I recently asked more than 70 eminent researchers if they would have done their work differently if they had thought Darwin's theory was wrong. The responses were all the same: No.

I also examined the outstanding biodiscoveries of the past century: the discovery of the double helix; the characterization of the ribosome; the mapping of genomes; research on medications and drug reactions; improvements in food production and sanitation; the development of new surgeries; and others. I even queried biologists working in areas where one would expect the Darwinian paradigm to have most benefited research, such as the emergence of resistance to antibiotics and pesticides. Here, as elsewhere, I found that Darwin's theory had provided no discernible guidance, but was brought in, after the breakthroughs, as an interesting narrative gloss.

In the peer-reviewed literature, the word "evolution" often occurs as a sort of coda to academic papers in experimental biology. Is the term integral or superfluous to the substance of these papers? To find out, I substituted for "evolution" some other word— "Buddhism," "Aztec cosmology," or even "creationism." I found that the substitution never touched the paper's core. This did not surprise me. From my conversations with leading researchers it had become clear that modern experimental biology gains its strength from the availability of new instruments and methodologies, not from an immersion in historical biology.[5]

It is not just scientific research that suffers when it is straightjacketed by evolutionary dogma. Science education is also negatively affected, and shockingly so. The teaching of embryology, for example, has been damaged by the evolutionary paradigm to an embarrassing degree. In 1874 Ernst Haeckel published drawings of a handful of species at different stages of embryological development. The drawings gave the impression that readers were witnessing something akin to time-lapse pictures of evolution in action—or to be more precise, pictures of embryological development recapitulating the evolutionary history

as it moved from the more primitive species to the more advanced—
ontogeny recapitulating phylogeny, as the saying goes. But the drawings, as it turned out, were fraudulent. Haeckel had minimized major
differences among the pictured animals at the earliest developmental
stages. It is well known in academia that these diagrams are fudged.[6]
Yet they and similar reproductions have been widely incorporated into
major biology textbooks ever since.[7] Whole generations of biology students thus have been fed misinformation about embryology.

Or to take another example, biology classes on the human ear
often include stories of how the ear supposedly evolved from reptiles.
Instead of such wild speculations, students would be far better off
learning about the advanced acoustic engineering of the ear.

The Intelligent Design Paradigm Is Good for Science

What about intelligent design? Anyone tuned into the evolution/
design controversy has heard the dire warnings about the grave dangers of allowing intelligent design into the sciences. The irony is that
the scientific revolution was led by men such as Copernicus, Galileo,
Boyle, Kepler, and Newton, all of whom owed much of their success
to the conviction that the world was the product of a rational intelligence. They reasoned that if the universe was intelligently designed,
then it made sense that scientists could investigate it and discover
designed laws.

If an archeologist were to discover a workshop belonging to Leonardo da Vinci, there would be a great effort to search for his designs,
and the investigators would be searching for brilliant designs, not
shoddy ones, since da Vinci is an acknowledged genius. Similarly,
the pioneers of the scientific revolution reasoned that if the universe
is the work of an ingenious designer, as they believed, then there is
bound to be sophistication and fine-tuning in nature. Their design
perspective gave them confidence that there were layers of rational
order to be discovered in nature. They went looking for that hidden
order, and they found it.

In the early seventeenth century Johannes Kepler was inspired
by his design perspective to search for the rational laws of planetary

motion. He discovered them and published the results in 1619. This was groundbreaking scientific work, which then encouraged others, including Isaac Newton, to study the physics of the universe. Newton made perhaps the biggest scientific breakthrough of all time when he formulated his laws of gravity and calculus. This laid much of the foundation for modern science.

Without a theistic perspective, the founders of modern science— Kepler, Newton, Boyle, Faraday, Maxwell, Fleming, and Kelvin, among others—would have lacked a key motivation in their scientific work. Sadly, the crucial role that Christian theism played in the thought of these pioneering scientists[8] typically goes unacknowledged in our Western education system. What is also not publicized is that great scientists such as Maxwell, Kelvin, and Fleming came after Darwin and did not convert to Darwinism.

The great scientists of the scientific revolution could see that intelligent design is entirely consistent with science. One of the most fundamental scientific laws is the first law of thermodynamics (conservation of energy). This law states that energy cannot be created or destroyed by natural processes; there can only be a change of form of energy. The obvious conclusion is that the universe requires a power beyond the universe to create energy in the first place. By ruling out intelligent design, modern science is preventing explanations that would be entirely consistent with scientific laws.

The intelligent design perspective, while distinct from theism, is both encouraged by, and encourages, an openness to theism. And this is good for science, because a theistic perspective—particularly Judeo-Christian theism—anticipates and expects advanced and sophisticated design in nature. This expectation encourages one to search out the purpose of various mysterious biological features rather than throwing up one's hands and assuming bad design.

How the Design Paradigm Fueled My Research Success

What was true at the birth of science in Christian Europe remains true today. I am living proof of this. I have had the honor of receiving many national awards for engineering and bioengineering design in the UK. I can state categorically that my intelligent design framework was crucial to my success. If I had followed the evolutionary paradigm, I doubt I would have won any of those prizes, because it would have stifled my research. I have met many researchers around the world, including biologists, who have also been inspired by an intelligent design perspective and owe their success to it.

My ID perspective led me to anticipate that the human foot would have a very sophisticated arched design superior to any human prosthetic, and this was found to be correct. I went on to develop an advanced bioinspired arched robotic foot. Had I believed Nathan Lents and Jeremy DeSilva that the human foot is a poor design, I would have been put off even investigating the human foot and lost that research opportunity.

My intelligent design perspective also led me to think the knee likely had a sophisticated linkage design, and that too proved correct. I went on to spend twenty-five years researching advanced bioinspired knees for robotics, overseeing PhD students in the work and funded by various grants. If I had believed Lents that the knee is a bad design, I might never have bothered to investigate the knee joint and would have missed this great research opportunity.

My design perspective led me to realize that multifunctioning in biological systems, such as the human throat, far from being ill-considered, allows for more compact designs that afford numerous benefits. This inspired me to write a review paper showing how engineers would do well to copy the multifunctioning strategy found throughout biology. If I had listened to Abby Hafer, who insisted that the human throat is a bad design, I might have missed this important research opportunity.

My work on dragonfly flight is yet another area where my intelligent design perspective stood me in good stead. According to evolutionary theory, dragonflies are relatively primitive creatures, having arrived so early on the scene. However, I studied them expecting to find sophisticated design, and my research benefited from that expectation. My work on dragonflies led to the development of a unique design of a microdrone that was reported in *New Scientist*, the popular international science magazine.[9]

Why Intelligent Design Is Excluded

I have met many scientists who have admitted that the cold-shouldering of intelligent design is primarily driven not by any scientific reason. In discussing the issue with my colleagues, they have given various explanations as to why they think ID is currently viewed as off-limits by so many scientists, particularly in biology. These reasons can be grouped into five main categories:

(1) Academic fear: not wanting to risk your career

A favorite quip in the intelligent design community was introduced by Jun-Yuan Chen, a Chinese paleontologist and acknowledged expert on the Cambrian explosion: "In China we can criticize Darwin, but not the government. In America, you can criticize the government, but not Darwin."[10] Funny but also sad—sad because science is meant to be about open debate and the unfettered testing of ideas. Its leaders are not supposed to act like despotic religious figures bent on silencing heretics.

The fear of losing one's career for questioning evolutionary theory is not an idle one. Some have lost research contracts or promotions for questioning the evolutionary paradigm. I know senior biologists who have lost academic positions for holding to an ID position. One can find examples easily enough, including in the documentaries *Expelled* and *Revolutionary*, and in the book *Canceled Science.*

Those who do not have their careers torpedoed can look forward to verbal abuse, as for instance when biochemist Larry Moran refers to design proponents as "IDiots," or when Richard Dawkins informs the world in stentorian tones that "it is absolutely safe to say that if you meet somebody who claims not to believe in evolution, that person is ignorant, stupid, or insane (or wicked, but I'd rather not consider that)."[11]

Such mudslinging serves to warn academics not to question the status quo, at least not publicly. Many, it seems, have received the message. I know quite a few biologists who have confessed to keeping their belief in intelligent design a secret.

(2) Academic laziness: too much hassle to question evolution

Most of the academics I have rubbed shoulders with in the UK and Europe would characterize themselves not as atheists but as agnostics, with no strong views about intelligent design. Many of these men and women have admitted that the evolutionary paradigm could well be wrong, but have concluded that it would be too costly and too much hassle to question and change an ingrained perspective. Some of them have, with remarkable honesty, told me that the most important things to them are the three p's: their pay, their pension, and their parking space. The fact that they are teaching a false theory of origins is, for them, not the most important thing in their list of priorities.

(3) A God allergy: disliking the repercussions of there being a Designer

Some academics who dislike the theory of intelligent design seem to do so because of non-scientific issues. If ID were acknowledged as true, then one would be prompted to ask, Who is the designing intelligence behind it all, and if the designer is God, are there absolute moral standards after all? Or to come at it from the other side, I have met scientists who have frankly admitted that they support evolutionary theory not primarily for any scientific reason but because they do not want God in the picture.

That is by no means a necessary motivation for embracing Darwinism, but I would argue that it is a sufficient one for many. An anti-God agenda is suggested by how very many vociferously anti-Christian thinkers took to the theory from the start. One of Darwin's earliest champions was Thomas Huxley. He styled himself an agnostic, but even many of his allies understood this to be a politically convenient cover for a man with a thorough-going materialist philosophy of life. Huxley didn't actually buy Darwin's insistence that natural selection drove the evolutionary process, but he recognized that Darwin's theory was the most cogent design-defeater on offer at the time, so in public Huxley backed it to the hilt. He was also a key organizer of the X-Club, dedicated to promoting scientific naturalism and pushing religious thinking to the margins of society.

Thomas Huxley's grandson, Sir Julian Huxley, was another prominent promoter of evolutionary theory. Himself an anti-religious agnostic, he became the first president of the British Humanist Society. More recently, the prominent evolutionists and dedicated ID critics Richard Dawkins and Alice Roberts have been vice presidents of Humanists UK, a society committed to an anti-theistic humanism.

Also, Dawkins along with Christopher Hitchens, Sam Harris, and Daniel Dennett, "the four horsemen" of the New Atheist movement, all promoted modern evolutionary theory in their case against God and Christianity. Their attitude is understandable. They freely confess to disliking the idea of an all-knowing Creator, and as Dawkins succinctly put it, "Darwin made it possible to be an intellectually fulfilled atheist."[12]

I should note that atheists of the militant "four horsemen" variety are the exception (although they have gotten a disproportionately large amount of airtime in the media). My sense from rubbing shoulders with numerous scientists over the years is that most who have a God allergy find the atheistic theatrics of the four horsemen off-putting and would prefer simply not to think about the possibility of God, much less talk about him at every turn. And for these more genial religious skeptics, a handy way for them to steer around the matter is simply to mark the theory of intelligent design as outside the bounds of proper science.

(4) Career inertia: decades of research invested in the evolutionary paradigm

Some scientists, including many paleontologists and most paleobiologists, have a long research career invested in defending macroevolution. Such academics have a strong vested interest in the status quo.

This was pressed home to me when I met up at a cafe with two well-known paleontologists who tried to persuade me not to write in support of intelligent design. During our (at times emotional) discussion it became obvious to me that their entire careers were built on the evolutionary paradigm. The idea that all their academic work might have been misguided was not something they wished to contemplate.

(5) Paradigm weakness: the theory of evolution is too vulnerable

Some of my academic colleagues have admitted to me that one reason ID is debarred is that the evidence for it is too strong. Particularly as regards the origin of life, my colleagues have stated that if intelligent design were allowed into the conversation of possible explanations, defending a fully naturalistic hypothesis for life's origin would simply be too difficult. The solution for those wedded to evolutionary naturalism is to keep intelligent design on the outs. The strategy is not unlike that of a dictator who, displeased by his poll numbers, bans the opposition politician from standing for election.

The Elephant in the Room

Given that the intelligent-design paradigm better explains the ultimate engineering we find in biology, surely it is past time for the ID perspective to have a place at the table.

In a sense it already does—or at least, it is in the room. What I mean is that many biologists have told me over the years that intelligent design is the proverbial elephant in the room. They mean that every biologist can see astonishing design sophistication in the complex systems of biology, but no one is allowed to discuss the possibility of purposive design.

At the end of the day, however, the issue isn't a polling question whose true answer is determined by counting the heads of those deemed fit to vote on the matter. In our investigation of biological origins, what we are trying to ascertain cannot be answered by determining whether there is or isn't a consensus on the question of biological origins. There was once a consensus that the continents do not move. Eventually that consensus was swept away by reality and replaced by the plate tectonics framework. Both before and after the earlier consensus, the continents moved. Science is not a search for the *consensus* about the natural world. It is a search for the *truth* about the natural world. And if we remain committed to following the evidence wherever it leads, the truth will out.

Actually, I have faith that the truth will out regardless—eventually.

For many of us, it already has. For my part, as an engineer who has learned to spot the signature of engineering genius in human inventions, I can see hallmarks of intelligent design practically everywhere I look in the living world.

22. My Experience Advocating Intelligent Design

Two Surprises

I have noted the pattern of evolutionists attacking design proponents in academia. That pattern is all too real. But it isn't the whole story, and I do not want to leave readers with a distorted picture.

I have been very public in my support for intelligent design, dating back to my year 2000 book *Hallmarks of Design*. In the intervening years I have been surprised by the level of pushback from some evolutionists, but I have been equally surprised by the level of support from people like Prince Charles (now King Charles) and from many academics. It has been humbling how many non-religious colleagues have supported me in private and in public. I will share a few experiences.

Intelligent Design in the Coffee Room

I have discussed the question of origins with many senior academics during coffee room chats and been surprised by how many are at least open to the concept of intelligent design. In most cases, these academics will not so much as hint at their openness in public, but there are exceptions. When I was at Cambridge University, I published research papers with Mike Ashby, one of the most eminent engineering scientists in the UK and a Fellow of the Royal Society.

In the introduction to one of his books on mechanical design, he wrote the following: "Nature, to some, is Divine Design; to others it is design by Natural Selection."[1]

Ashby wrote those words because during his time at Cambridge, he discovered that there were a significant number of academics who supported intelligent design. While at Cambridge, I shared publications on intelligent design with him and several other academics in the engineering department, and received very supportive feedback. I have found that engineering academics are often more ready than biologists to acknowledge that intelligent design is a fully scientific concept, because they understand the difficulty of designing complex systems.

Intelligent Design in a Lunchtime Lecture

In 2006 I gave a lunchtime intelligent design talk at my university, which turned out to be bigger and more controversial than planned. It was initially meant to be a low-key lecture to a small group of around ten pro-ID students. However, one undergraduate managed to find email lists that enabled him to invite the entire university. And he did not tell me!

When the venue was changed to the largest lecture room in the engineering department, I could not understand why. But when I turned up, the reason became clear. When I entered the large lecture room, I was surprised to see hundreds of students and lecturers, including many senior professors from the biology and paleontology departments.

I spent forty-five minutes summarizing evidence for intelligent design and an hour in Q & A. It became clear that many students were fascinated by my argument that the theory of intelligent design is supported by strong scientific evidence. But it was also clear that a minority of academics were angry I was giving a pro-ID lecture within the university.

The following day I was passing my dean, and I asked him if he had heard about the lecture, to which he replied, "The whole university is discussing that ID lecture!" He also told me he strongly supported my right to free speech.

Following the lecture, I was surprised how many academic colleagues came to me and said they also took the ID position or were sympathetic to it. Most of them (especially the biologists) did not want to go public about it. In one case a senior biologist suggested he join me in teaching a unit that critiqued the theory of evolution. However, after a few weeks he said he realized it could cost him his career, and he pulled out. It was sad seeing firsthand how academic bullying can damage science and education.

Many students also came to me and said they really wished there could be more debates about intelligent design within the university. They were disappointed when the university decided that one ID lecture was enough.

I received several friendly emails from students after the lecture. Here are a few:

"The biology department asked people to… oppose your point of view and questions had been pre-arranged—an attitude I find strange as a student."
—Biology student

"The talk was amazing—people being turned away as we packed out the lecture theatre… Thank you for the tremendous gentleness you showed despite a lot of angry biologists."
—Engineering student

"I have never seen a lecture have such a big impact on so many people."
—Biology student

"Really enjoyed the lecture and Q and A. (I spent most of the Q and A trying to understand what was being said!)"
—Engineering student

I found this last comment, from a student "trying to understand what was being said," quite funny. His difficulty stemmed not from any lack of intelligence but from his being completely new to the ID/ evolution conversation. There is a serious point here. Many students are unaware of the extensive arguments against evolutionary theory and for intelligent design, the educational system up to that point having studiously shielded them from any such dangerous information.

Intelligent Design in the Classroom

I have always been careful not to promote my belief in intelligent design in my university lectures. This is despite the fact that people such as Richard Dawkins and Alice Roberts have explicitly promoted their atheistic worldview in their university lectures. However, when students have asked me questions about evolution or origins, I have given factual answers.

For example, engineers sometimes use what are known as "genetic algorithms" to search for optimal or near-optimal solutions to engineering problems. These are computer programs partially modeled on Darwin's evolutionary mechanism of random variation and natural selection. Students have asked me whether these genetic algorithms prove that evolution works in nature. I explain that these algorithms are based on adaptation, a well-established fact of science. I also explain that while mutations are included in these computer-based algorithms, each such mutation involves multiple simultaneous parameter changes, which is quite different from the small, random genetic mutations we find contributing to microevolutionary adaptation in nature and that are said to drive the Darwinian process. Later evolutionists, following Darwin, rejected "hopeful monster" mutations, with their many simultaneous changes, as being too ridiculously improbable, and they did so for good reason. Beneficial big-jump mutations have been ruled out by more than a century of mutagenesis experiments.[2]

So why do the genetic algorithms employed by engineers employ big mutations when big-jump mutations have been ruled out by so much laboratory evidence? The computer programs diverge from biological reality in this way because otherwise the programs would fail to evolve anything useful. Also of note, the genetic algorithms employed by engineers are intelligently designed. For these two reasons, the genetic algorithms used in engineering actually provide evidence against rather than for evolutionary theory.

Additionally, researchers William Dembski, Robert Marks, and Winston Ewert demonstrated that the algorithms developed to demonstrate and explore the creative power of biological evolution can only generate nontrivial results if they are provided with information about the desired results.[3] This is damning because a central tenet of mainstream evolutionary theory is that no such information is provided by nature, since it lacks foresight.

On one occasion I was asked whether the human back was a bad design due to its having evolved from knuckle-walking ape-like ancestors. I answered that the available scientific studies show the human back to be a very good design. I also noted that human origins is a controversial subject with many different perspectives and is complicated by the question being of great significance to various worldviews. Following this exchange, I received an email from an editor of the student campus newspaper. "I am currently writing an article with regards to a complaint made against you to the Dean," the student informed me. He explained that "there have been accusations made that you often stress your religious views within the normal lecture program."[4]

This was a rather worrying email to receive. However, when I explained to the editor what had really occurred in my lectures, he was understanding and withdrew the article. He realized that the accusations were exaggerated and unfair. The fact that a complaint had been made to the dean was not news to me, because my dean had already told me. The dean was not religious himself but was also very supportive of my position.

Intelligent Design in the National Media

On one occasion, an accusation about my teaching made it into a national newspaper. I was having coffee at my university and saw one of my colleagues reading the *Times Higher Education Supplement* (June 23, 2006). The front-page headline read "Intelligent Design Creeps on to Courses."[5] I joked with my colleague that someone must be in trouble. His reply was: "Yes, and it is you! Because your picture is inside and there is a major article claiming that you are teaching ID in your courses!" Well, that was a surprise to me. That article, unsurprisingly, raised quite a few eyebrows at my university, but I was not sanctioned, because my colleagues knew that if I ever spoke about intelligent design, it was only in response to student questions.

On another occasion I was interviewed live on the BBC after release of the science documentary *Unlocking the Mystery of Life*. Before the interview I was told that the BBC wanted me to explain to viewers about how the documentary supports intelligent design. I was surprised when the first question was, "Who is the intelligent designer?" My first thought was, that should be the last question, not the first! Sensing that the interviewer was trying to trip me up, I merely noted that the intelligent designer was clearly extremely good at design and proceeded to give examples of brilliant design in biology.

Not all journalists are anti-ID. A few months after the *Times Higher Education Supplement* story, another national newspaper, *The Independent*, invited me to write an article in support of intelligent design. The article was published in February 2007 and was entitled "Against the Grain: 'There Are Strong Indications of Intelligent Design,'" with a short summary paragraph that included the sentence, "He argues that intelligent design is a valid scientific theory."[6]

Intelligent Design at Engineering Conferences

I once gave a keynote lecture at a *Design and Nature* international conference organized by the Wessex Institute of Science and Technology, based in the UK. My talk was about my work on bioinspired drones based on the design of dragonflies. The talk before mine was by a speaker who said he had evolved his design using genetic algorithms inspired by Darwinian evolution. At the start of my talk, I briefly mentioned that while his genetic algorithms accurately mimic biological microevolution, they did not mimic macroevolution because they involved many simultaneously planned mutations.

My remarks sparked quite a debate, and I was struck by how many engineering researchers quickly saw how genetic algorithms do not necessarily support evolutionary theory. In fact, the very next speaker at the conference, who was from Turkey, started his presentation by stating that he agreed with my comments about genetic algorithms, and he stated that he was an advocate of intelligent design. This is another example of how, when intelligent design is allowed to be discussed in public, you suddenly find there is significant support for ID. This may explain why the Dawkinses of this world have worked so hard to strictly limit or ban discussion of intelligent design in academic settings.

Support from Sir Michael Berry

In 2006 an organization called the British Centre for Science Education (BCSE) wrote to world-famous physicist Sir Michael Berry, telling him I had said that I thought that he worked for the Devil. Of course, I had done no such thing. The BCSE, I should mention, has nothing to do with serious science. Instead, it is a group of anti-ID fanatics dedicated to attacking anyone who believes in intelligent design. The reason they named the organization the British Centre for Science Education is to mislead the public into thinking they have weighty scientific credentials.

I did not know that the BCSE were trying to stir up trouble for me until I received an email from Berry. When I read it I was shocked but also grateful that Berry had done the stand-up thing and checked with me directly rather than assuming the charge was true. When I explained the reality of the situation and the background of the BCSE, he was very sympathetic.

The incident led to a friendly discussion between the two of us about worldviews. He explained that he was an atheist but respected my ID perspective. Following the scurrilous action by the BCSE, I was honored by his attending my inaugural professorial lecture in the engineering department, a lecture to celebrate my becoming a full professor. After the lecture he wrote to me, saying, "I enjoyed your lecture very much in style and content."

Support from My Head of School

I have collaborated with biologists at many points in my career, and in the vast majority of cases they have had no difficulty with my intelligent design perspective. However, there was one occasion when the head of the engineering school informed me that I was being excluded from a biology research project simply because of my views on evolution and intelligent design. Although he was not an ID proponent, he told me he thought this was outrageous and would fully support me making an appeal. In this case I was happy to let the project go because I had enough projects to work on. But I appreciated his support.

I once experienced intense opposition from a humanist group after giving a pro-design lecture in Winchester Town Hall in Britain. The group published a critique of my lecture online and sent the link to senior academics at my university to try to stir up trouble. Their efforts backfired because my head of school wrote back with a very comprehensive message of support. A short excerpt from his reply is given here:

> I know Stuart Burgess; he is a colleague…. I am NOT an intelligent design believer, nor a Christian…. Professor Burgess is a professor of design who teaches and researches design within Mechanical Engineering (his mechanical insect for example, is very very clever). Contrast this with Richard Dawkins who is a professor of "public understanding of science" i.e. "Pushing his opinions on the public." I personally find it deeply offensive that Dawkins's "God delusion" opinion should be called science, whereas Stuart Burgess's opinion should be called "dangerous" or "laughable."[7]

The above example is interesting because I have found that most academics, at least in the UK, are agnostic and prefer not to engage in debates about origins. However, when they witness irrational attacks on ID, it often spurs them to step forward and vocally support the free-speech rights of pro-ID academics.

Support from University Colleagues

My university has been very supportive of my right to freedom of speech, senior engineering professors in particular. I do wonder if their expert engineering knowledge played a part in their choosing to support someone with an ID perspective. That being said, I also have had support from several biologists at the university.

I remember one interesting encounter when I was being interviewed by a high-level university committee to assess my engineering department. At the time I was chair of the department, responsible for six hundred students and a multimillion-dollar budget. To break the ice, the committee chairman, a senior biologist, joked, "Well, this is the guy who is questioning one of the central paradigms in biology!" At that point, everyone smiled. He had succeeded in breaking the ice. I appreciated the friendly recognition of my perspective on origins science.

Several anti-ID zealots have written to my university urging them to force me to put a disclaimer in my pro-ID books and articles that my views are not connected with my work at the university. This would be ridiculous, because my writings arguing for intelligent design are based mainly on the engineering research I have carried out at my university. However, some people feel that intelligent design should be banned no matter what the evidence shows.

Excerpts from one such letter are below. It was written by a freelance journalist for the *Belfast Telegraph*, Lewis Read. Five key parts of his letter are shown, followed by responses from the university, which were originally given in all-caps, as below.[8]

"Dear Sirs

I attended Professor Burgess's talk: The Design Argument. He spoke for perhaps 90 minutes. What struck me was that for a large portion of his talk he laboured heavily the fact that he is a professor at Bristol University, bedazzling us in the audience with his work on the ENVISAT satellite. After much performance with diagrams and algebraic equations he argued for intelligent design in nature. I bought his book, *Hallmarks of Design,* and under acknowledgements Professor Burgess signs off with his address at Bristol University. The book is also endorsed by Professor Alan Linton formerly Head of Microbiology at Bristol University.

My questions are
    Is the university aware that Professor Burgess is giving this kind of advertising on his intelligent design writing?
    Does the university endorse this material?
    Does the university approve of its name being attached to the creation books by Burgess?

If it does endorse this material, on what scientifically accepted principles?

If it does not endorse this material, what action does it intend to take to ensure its name is not used in this way?

I interviewed people at the talk who seemed entirely convinced by what he said. What is your message to anyone who was perhaps persuaded by him on the basis of his authority as a professor at Bristol University?

Question: Is the university aware that Professor Burgess is [advertising the university name] in his ID writing?

Answer: WE DO NOT REGARD THIS AS "ADVERTISING." IT IS A STATEMENT OF FACT—PROFESSOR BURGESS DOES INDEED WORK AT THIS UNIVERSITY.

Question: Does the university approve of its name being attached to this publication?

Answer: IT WOULD BE DIFFICULT TO OBJECT TO A STATEMENT OF FACT—AS I SAID EARLIER, PROFESSOR BURGESS DOES WORK HERE.

Question: If it does endorse this material, on what scientifically accepted principles?

Answer: N/A SEE EARLIER RESPONSE

Question: If it does not endorse this material what action does it intend to take to ensure its name is not used in this way.

Answer: N/A SEE EARLIER RESPONSE

Question: I interviewed people at the talk who seemed entirely convinced by what he said. What is your message to anyone who was perhaps persuaded by him on the basis of his authority as a professor at Bristol University?

Answer: UNIVERSITIES GENERALLY PROMOTE FREEDOM OF THOUGHT AND EXPRESSION. IT IS FOR PEOPLE TO MAKE UP THEIR OWN MINDS ABOUT SUCH MATTERS, NOT FOR THE UNIVERSITY TO SUGGEST WHAT THEY SHOULD OR SHOULD NOT THINK.

It is important to note that atheists such as Richard Dawkins, Jerry Coyne, and Nathan Lents often mention their academic credentials when arguing from their atheistic perspective. It would hardly be fair to deny intelligent design proponents the right to do the same when arguing for design and purpose in the universe.

The Thought Police

If the attacks described above on my academic freedom are not Orwellian enough, consider this: I have had complaints leveled against me that because some of my scientific publications support intelligent design, I must have spent time in my academic job thinking about intelligent design. The complainants were convinced that my academic position at the university did not allow me to consider, at least while on the clock, whether there was evidence for design in nature. This is despite the fact that my research was focused on design in nature and I was a co-editor of the journal *Design in Nature*.

One sample complaint from a university science research director was made to the research director of my engineering school. His complaint stated, "The concern with Professor Burgess is that he is using his work address for publications on intelligent design that are not linked to what he is employed to do." Yes, I was being accused of a thought crime. Big Brother would surely approve.

We sometimes hear the false claim that intelligent design proponents do not publish scientific work supporting ID. Well, given that there is pressure on us to not even think about the possibility of intelligent design, it would hardly be surprising if pro-ID researchers were not eager to stick their necks out by publishing pro design scientific papers. The surprising thing is how many such peer-reviewed pro-ID papers have survived both the peer-pressure and the peer review process and seen the light of day over the past two decades—more than two hundred and rising as of a published report issued in May 2024.[9]

Threatening Emails

Sometimes evolutionists have written directly to me challenging my stance. In such cases I always try to engage with their arguments and concerns. There have been many instances where I have had a very cordial and friendly exchange. However, some emails have been quite threatening. An excerpt from one such email is given here:

> Dear Stuart,
>
> I have heard that you welcome discussions with your students about design in nature. We would regard such a situation as being very serious and would seek to ensure that you were called to account. As always, we wish to be fair to you before proceeding to the next stage.
>
> The Brights Atheist Association, email, 13 Nov 2003

When I received this email I was quite surprised that anyone would object to me talking to students about design in nature. In fact, at the time I was actually teaching a course entitled, "Design in Nature," so it would have been difficult not to talk students about this subject. However, when I read the line about "proceeding to the next stage," I did wonder, What was the next stage?

Dawkins's Admission

I once had a brief debate with Richard Dawkins on the subject of intelligent design. It played out on the letters page of a UK national newspaper, *The Guardian*. I (together with one of my colleagues, Andy McIntosh) explained how scientific principles like the second law of thermodynamics support intelligent design. Dawkins countered by arguing that intelligent design proponents were a tiny minority in academia and took a position radically at odds with the evolutionary paradigm:

> Maybe Burgess and McIntosh are right and all the rest of us—biologists, geologists, archaeologists, historians, chemists, physicists, cosmologists and, yes, thermodynamicists and respectable theologians, the vast majority of Nobel Prize winners, Fellows of the Royal Society and of the National Academies of the world—are wrong. Not just slightly wrong but catastrophically, appallingly, devastatingly wrong.

... My purpose in this article is to convey the full magnitude of the error into which, if Burgess and McIntosh are right, the scientific establishment has fallen.[10]

When Dawkins writes, "Maybe Burgess and McIntosh are right," I agree! However, it is not the scientific establishment that is in danger of falling but the Darwinian paradigm. And Dawkins is mistaken to assume that academics support the paradigm of designer-free evolution with almost perfect uniformity. My more than thirty years of academic experience tells me that there is widespread lack of confidence in the idea and a growing openness to intelligent design.

You Don't Need a PhD

Like many other researchers today, I am awestruck by the ultimate engineering of biological systems and can see how this points to intentional and purposeful design. But you do not have to have research grants or a PhD to be awestruck by design in biology. You only need to look at your own hands and feet, or consider your creative abilities, or contemplate your capacity to sense beauty. If there is an intelligent designer, people ask, shouldn't he demonstrate his existence? Well, I believe the designer has done so, through ultimate engineering, through extremes of design, and most of all, through the design of humankind.

ENDNOTES

INTRODUCTION

1. Andrea Rinaldi, "Naturally Better: Science and Technology Are Looking to Nature's Successful Designs for Inspiration," *EMBO Reports* 8 (2007): 995, https://doi.org/10.1038/sj.embor.7401107.

2. Steven Vogel, *Cats' Paws and Catapults: Mechanical Worlds of Nature and People* (New York: W. W. Norton, 1998), 23.

3. François Jacob, "Evolution and Tinkering," *Science* 196, no. 4295 (June 10, 1977), 1163–1164, https://www.science.org/doi/10.1126/science.860134. Internal reference removed.

4. Nathan H. Lents, *Human Errors: A Panorama of Our Glitches, from Pointless Bones to Broken Genes* (New York: Houghton Mifflin Harcourt, 2018).

5. Abby Hafer, *The Not-So-Intelligent Designer: Why Evolution Explains the Human Body and Intelligent Design Does Not* (Eugene, OR: Cascade Books, 2015), 110, ref. 6.

6. Richard Dawkins, *The Greatest Show on Earth: The Evidence for Evolution* (London: Black Swan, 2010), 354, 371, and 356 respectively.

7. Andrea Rinaldi, "Naturally Better," 995.

8. Examples of pre-programmed adaptive capabilities include epigenetic changes, which are programmed changes that enable an organism to adapt itself or its behaviors to changes in its ecosystem. It has been shown that organisms can inherit such adaptive modifications in their DNA in the form of chemical tags that influence how genes express physical traits. Epigenetics is a growing field of research, and epigenetic processes have all the hallmarks of being an intelligently designed process.

9. Theodosius Dobzhansky, *Genetics and the Origin of Species* (New York: Columbia UP, 1937), xxvii–xxviii.

10. Charles, Prince of Wales, "A Royal View," *BBC Radio 4 Reith Lectures*, 2000, http://news.bbc.co.uk/hi/english/static/events/reith_2000/lecture6.stm.

CHAPTER 1: FOOT AND ANKLE

1. Stuart Burgess et al., "A Bio-Inspired Arched Foot with Individual Toe Joints and Plantar Fascia," *Biomimetics* 8, no. 6, 455 (2023), https://doi.org/10.3390/biomimetics8060455.

2. Salih Angin and İlkşan Demirbüken, "Ankle and Foot Complex," in *Comparative Kinesiology of the Human Body: Normal and Pathological Conditions*, eds. Salih Angin and İbrahim Engin Şimşek (London: Elsevier, 2020), 411–439.

3. Amaraporn Boonpratatong and Lei Ren, "The Human Ankle-Foot Complex as a Multi-Configurable Mechanism During the Stance Phase of Walking," *Journal of Bionics Engineering* 7 (2010): 211–218, https://doi.org/10.1016/S1672-6529 (10)60243-0.

4. Claire L. Brockett and Graham J. Chapman, "Biomechanics of the Ankle," *Journal of Orthopaedic Trauma* 30, no. 3 (2016): 232–238, https://doi.org /10.1016/j.mporth.2016.04.015.

5. M. Bozkurt et al., "Functional Anatomy of the Ankle," in *Sports Injuries*, ed. Mahmut Nedim Doral and Jon Karlsson (Berlin: Springer, 2014), 1744.

6. "Imaging of the Forefoot and Midfoot," *Radiology Key*, January 17, 2016, radiologykey.com/imaging-of-the-forefoot-and-midfoot/, https://radiologykey .com/imaging-of-the-forefoot-and-midfoot/.

7. Nathan H. Lents, *Human Errors: A Panorama of Our Glitches, from Pointless Bones to Broken Genes* (New York: Houghton Mifflin Harcourt, 2018), 29.

8. Jeremy DeSilva, "Starting Off on the Wrong Foot: How Our Ape Ancestry Predisposes Us to Foot and Ankle Maladies," (lecture, AAAS annual meeting, Boston, MA, February 14–18, 2013).

9. The Scars of Evolution: AAAS Annual Meeting, Boston, MA, February 14–18, 2013, https://aaas.confex.com/aaas/2013/webprogram/Session5714.html.

10. Jeremy DeSilva, "Starting Off on the Wrong Foot."

11. Boonpratatong and Ren, "The Human Ankle-Foot Complex as a Multi-Configurable Mechanism During the Stance Phase of Walking."

12. Madhusudan Venkadesan et al., "Stiffness of the Human Foot and Evolution of the Transverse Arch," *Nature* 579 (2020): 97, https://doi.org/10.1038/s41586 -020-2053-y.

13. B. Baldisserri and V. P. Castelli, "Passive Motion Modeling of the Human Ankle Complex Joint," in *Proceedings of 13th World Congress in Mechanism and Machine Science* (2011): 1–7, https://www.dmg-lib.org/dmglib/streambook/index .jsp?bookid=22568009.

14. Stuart Burgess, "Why the Foot Is a Masterpiece of Engineering and a Rebuttal of 'Bad Design' Arguments," *Biocomplexity* 2022, no. 2 (November 2022): 1–10, https://bio-complexity.org/ojs/index.php/main/article/view/BIO-C.2022.3 /BIO-C.2022.3. Note that this chapter benefits at several points from material in this 2022 paper.

15. Burgess, "Why the Foot Is a Masterpiece of Engineering."

16. Burgess et al., "A Bio-Inspired Arched Foot with Individual Toe Joints and Plantar Fascia."

17. We do have wrist bones from Lucy, and they indicate that she may have knuckle walked. See Mark Collard and Leslie C. Aiello, "From Forelimbs to Two Legs," *Nature* 404 (March 23, 2000): 339–340, https://doi.org/10.1038/35006181.

18. Carol Ward, William H. Kimbel, and Donald C. Johanson, "Complete Fourth Metatarsal and Arches in the Foot of *Australopithecus afarensis*," *Science* 331, no. 6018 (2011): 750–753, https://www.science.org/doi/abs/10.1126 /science.1201463.

19. "Foot Bone Suggests Lucy's Kin Had Arched Foot, for Walking," *Science Daily*, February 10, 2011, https://www.sciencedaily.com/releases/2011/02/110210141213.htm; "New View of Human Evolution?," *Science Daily*, February 11, 2011, https://www.sciencedaily.com/releases/2011/02/110210141215.htm.

20. P. J. Mitchell, E. E. Sarmiento, and D. J. Meldrum, "The AL 333–160 Fourth Metatarsal from Hadar Compared to that of Humans, Great Apes, Baboons, and Proboscis Monkeys," *HOMO: Journal of Comparative Human Biology* 63, no. 5 (October 2012): 336, https://doi.org/10.1016/j.jchb.2012.08.001.

21. Lents, *Human Errors*, 29.

22. "Ankle Fusion (Arthrodesis)," The Royal Orthopaedic Hospital NHS Foundation Trust (Birmingham, UK), accessed October 11, 2025, https://roh.nhs.uk/services-information/foot-and-ankle/ankle-fusion.

23. Rohan Newman, "Ankle Fusion: Procedure, Recovery, and Long-Term Outlook," Newman Feet, 2025, https://newmanfeet.com/ankle-fusion-procedure-recovery-and-long-term-outlook/.

24. Htwe Zaw and James D. F. Calder, "Tarsal Coalitions," *Foot and Ankle Clinics* 15, no. 2 (June 2010): 349–64, https://www.foot.theclinics.com/article/S1083-7515(10)00022-7/abstract.

25. Lents, *Human Errors*, 30.

26. Eiichi Uchiyama et al., "Distal Fibular Length Needed for Ankle Stability," *Foot and Ankle International* 27, no. 3 (March 2006): 185, https://doi.org/10.1177/107110070602700306.

27. Q. Wang et al., "Fibula and its Ligaments in Load Transmission and Ankle Joint Stability," *Clinical Orthopaedics and Related Research* 330 (1996): 261–270, https://pubmed.ncbi.nlm.nih.gov/8804301/.

28. Lents, *Human Errors*, 29.

29. M. Walther et al., "Injuries and Response to Overload Stress in Running as a Sport," *Orthopade* 34, no. 5 (May 2005):399–404, https://pubmed.ncbi.nlm.nih.gov/15841366/; R.J . Walls et al., "Football Injuries of the Ankle: A Review of Injury Mechanisms, Diagnosis and Management," *World Journal of Orthopedics* 7, no. 1 (2016): 8–19, https://pubmed.ncbi.nlm.nih.gov/26807351/; Eric Giza et al., "Mechanisms of Foot and Ankle Injuries in Soccer," *American Journal of Sports Medicine* 31, no. 4 (2003): 550–554, https://pubmed.ncbi.nlm.nih.gov/12860543/.

30. Lents, *Human Errors*, 29.

31. Alice Roberts, "Can Science Make Me Perfect?," *BBC4*, June 13, 2018, audio, 1:29:00, https://www.bbc.co.uk/programmes/b0b6q3qy.

32. For an informative video on this fascinating subject, with animation of the molecular processes, see "The Robot Repairmen Inside You," *Secrets of the Cell*, episode 9, Discovery Institute, October 15, 2024, https://www.discovery.org/v/sotc9-robot-repairmen/.

CHAPTER 2: THE KNEE JOINT

1. S. D. Masouros et al., "Biomechanics of the Knee Joint," *Orthopaedics and Trauma* 24, no. 2 (April 2010): 84–91, https://doi.org/10.1016/j.mporth.2010.03.005.

2. A. C. Etoundi et al., "Bio-Inspired Knee Joint: Trends in the Hardware Systems Development," *Frontiers in Robotics and AI* 8 (2021): 613574.

3. Stuart C. Burgess and Appolinaire C. Etoundi, "Performance Maps for a Bio-Inspired Robotic Condylar Hinge Joint," *ASME Journal of Mechanical Design* 136, no. 11, 115002 (2014), https://asmedigitalcollection.asme.org/mechanicaldesign /issue/136/11; Appolinaire C. Etoundi, Stuart C. Burgess, and Ravi Vaidyanathan, "A Bio-Inspired Condylar Hinge for Robotic Limbs," *ASME Journal of Mechanical Design* 5, no. 3, 031011 (2013), https://doi.org/10.1115 /1.4024471; A. C. Etoundi, R. J. Lock, R. Vaidyanathan, and S. C. Burgess, "A Bio-Inspired Condylar Knee Joint for Knee Prosthetics," *International Journal of Design & Nature and Ecodynamics* 8, no. 3 (2013): 213–225, https://www.witpress .com/elibrary/dne-volumes/8/3/741; A. C. Etoundi, R. Vaidyanathan, and S. C. Burgess, "A Bio-Inspired Condylar Hinge for Mobile Robots," in *Proceedings of the International Conference on Intelligent Robots and Systems*, San Francisco, CA, September 25–30, 2011, 4042–4047, https://ieeexplore.ieee.org/document /6094924; S. C. Burgess and A. Etoundi, "A Bio-Inspired Limb Joint for Prosthetics," Published presentation, 6th International Conference on Design and Nature, June 2012, La Corina, Spain.

4. Simon Garrett, "Surprising Facts About Your Knees," *Hip and Knee Consultant*, accessed March 17, 2025, https://www.hipandkneeconsultant.co.uk/post /surprising-facts-about-your-knees.

5. Nathan H. Lents, *Human Errors: A Panorama of Our Glitches, from Pointless Bones to Broken Genes* (New York: Houghton Mifflin Harcourt, 2018), 21, 23.

6. This chapter benefits at several points from material drawn from my paper "How Multifunctioning Joints Produce Highly Agile Limbs in Animals with Lessons for Robotics," *Biomimetics* 9, no. 9 (2024): 529, https://www.mdpi. com/2313-7673/9/9/529.

7. Eleftherios A. Makris, Pasha Hadidi, and Kyriacos A. Athanasiou, "The Knee Meniscus: Structure-Function, Pathophysiology, Current Repair Techniques, and Prospects for Regeneration," *Biomaterials* 32, no. 30 (October 2011): 4, https://doi.org/10.1016/j.biomaterials.2011.06.037.

8. For more on the matter, see my paper "Universal Optimal Design in the Vertebrate Limb Pattern and Lessons for Bioinspired Design," Section 5.2, *Bioinspiration and Biomimetics* 19, no. 5 (September 2024), DOI 10.1088/1748-3190/ad66a3 and L. G. Hallén and O. Lindahl, "The 'Screw-Home' Movement in the Knee Joint," *Acta Orthopaedica Scandinavica*, 37, no. 1 (1966): 97–106, https://doi.org/10.3109/17453676608989407.

9. Burgess and Etoundi, "Performance Maps," Etoundi, Burgess, and Vaidyanathan, "A Bio-Inspired Condylar Hinge for Robotic Limbs," and Etoundi, Lock, Vaidyanathan, and Burgess, "Bio-Inspired Condylar Knee Joint for Knee Prosthetics."

10. A. G. Steele, A. Hunt, and A. C. Etoundi, "Development of a Bio-Inspired Knee Joint Mechanism for a Bipedal Robot," in *Biomimetic and Biohybrid Systems: 6th International Conference, Living Machines 2017*, eds. Michael Mangan et al. (Cham, Switzerland: Springer, 2017).

11. Takehito Kikuchi, Kohei Sakai, and Isao Abe, "Bioinspired Knee Joint for a Power-Assist Suit," *Journal of Robotics Special Issue: Biologically Inspired Robotics* (July 14, 2016), https://onlinelibrary.wiley.com/doi/full/10.1155/2016/3613715.

12. Burgess and Etoundi, "Performance Maps."

13. Lents, *Human Errors*, 22.
14. William J. Kraemer, Nicholas A. Ratamess, and Duncan N. French, "Resistance Training for Health and Performance," *Current Sports Medicine Reports* 1, no. 3 (2002): 165–171, doi: 10.1249/00149619-200206000-00007.
15. Callum McCaskie et al., "The Benefits to Bone Health in Children and Pre-School Children with Additional Exercise Interventions: A Systematic Review and Meta-Analysis," *Nutrients* 15, no. 1 (December 27, 2022): 127, doi: 10.3390/nu15010127.
16. "The Impact of Obesity on Bone and Joint Health," The American Academy of Orthopaedic Surgeons (AAOS), Position Statement 1184, 2015, https://www.aaos.org/contentassets/1cd7f41417ec4dd4b5c4c48532183b96/1184-the-impact-of-obesity-on-bone-and-joint-health1.pdf.
17. "The Impact of Obesity on Bone and Joint Health."

CHAPTER 3: THE WRIST JOINT

1. "7.10B: Muscles of the Wrist and Hand," *LibreTexts Health*, accessed October 13, 2025, search https://med.libretexts.org/ for "7.10B: Muscles of the wrist and hand."
2. Neil M. Bajaj, Adam J. Spiers, and Aaron M. Dollar, "State of the Art in Prosthetic Wrists: Commercial and Research Devices," IEEE International Conference on Rehabilitation Robotics (ICORR), Singapore (2015), 331–338, doi: 10.1109/ICORR.2015.7281221.
3. Stuart C. Burgess, "How Multifunctioning Joints Produce Highly Agile Limbs in Animals with Lessons for Robotics," *Biomimetics* 9, no. 9 (2024): 529, https://doi.org/10.3390/biomimetics9090529.
4. Nathan H. Lents, *Human Errors: A Panorama of Our Glitches, from Pointless Bones to Broken Genes* (New York: Houghton Mifflin Harcourt, 2018), 30.
5. For more on this and the biomechanics of the human wrist and hand, see C. Ayhan and E. Ayhan, "Kinesiology of the Wrist and the Hand," in *Comparative Kinesiology of the Human Body: Normal and Pathological Conditions*, eds. Salih Angin and Ibrahim Engin Simşek (London: Elsevier, 2022), 797–1130, https://archive.org/details/comparative-kinesiology-of-the-human-body-normal-and-pathological-conditions/page/n795/mode/2up.
6. Ayhan and Ayhan, "Kinesiology of the Wrist and the Hand," 900.
7. Lents, *Human Errors*, 28.

CHAPTER 4: FINGERS

1. T. M. W. Burton, R. Vaidyanathan, S. C. Burgess, A. J. Turton, and C. Melhuish, "Development of a Parametric Kinematic Model of the Human Hand and a Novel Robotic Exoskeleton," IEEE International Conference on Rehabilitation Robotics, Rehab Week, Zurich, Switzerland, June 29–July 1, 2011, doi: 10.1109/ICORR.2011.5975344.
2. Alice Roberts, *The Incredible Unlikeliness of Being: Evolution and the Making of Us* (London: Quercus, 2014), 171.
3. Michael Morgan and David Carrier, "Protective Buttressing of the Human Fist and the Evolution of Hominin Hands," *Journal of Experimental Biology* 216, no. 2 (2013): 236–244, doi: 10.1109/ICORR.2011.5975344.

4. R. W. Young, "Evolution of the Human Hand: The Role of Throwing and Clubbing," *Journal of Anatomy* 202, no. 1 (February 2003): 165–174, https://doi.org/10.1046/j.1469-7580.2003.00144.x.
5. Lisa Skedung et al., "Feeling Small: Exploring the Tactile Perception Limits," *Scientific Reports* 3 (September 12, 2013): 2617, https://doi.org/10.1038/srep02617.
6. Jacques Duchateau and Roger M. Enoka, "Distribution of Motor Unit Properties Across Human Muscles," *Journal of Applied Physiology* 132, no. 1 (January 2022): 1–13, https://doi.org/10.1152/japplphysiol.00290.2021.

CHAPTER 5: THE SPINE

1. S. C. Burgess, "Shape Factors and Material Indices for Dimensionally Constrained Shafts," *Proceedings of the Institution of Mechanical Engineers, Part C: Journal of Mechanical Engineering Science* 214 (2000): 381–388; "Shape Factors and Material Indices for Dimensionally Constrained Beams," *Proceedings of the Institution of Mechanical Engineers, Part C: Journal of Mechanical Engineering Science* 214 (2000): 371–379; "The Ranking of Efficiency of Structural Layouts Using Form Factors, Part II: Design for Strength," *Proceedings of the Institution of Mechanical Engineers, Part C: Journal of Mechanical Engineering Science* 212 (1998): 129–140; "The Ranking of Efficiency of Structural Layouts Using Form Factors, Part I: Design for Stiffness," *Proceedings of the Institution of Mechanical Engineers, Part C: Journal of Mechanical Engineering Science* 212 (1998): 117–128.
2. Jess Clark, "Human Spine Inspires Bridge Design Development," *New Civil Engineer*, February 1, 2018, https://www.newcivilengineer.com/archive/human-spine-inspires-bridge-design-development-01-02-2018/ .
3. Clark, "Human Spine Inspires Bridge Design Development."
4. Nathan H. Lents, *Human Errors: A Panorama of Our Glitches, from Pointless Bones to Broken Genes* (New York: Houghton Mifflin Harcourt, 2018), 25–27.
5. Some authors refer to a J-shaped human spine. However, this is essentially a slim S-shape. When a person is fit and slim they tend to have a slim S-shape.
6. Katherine J. Wu, "Hero Shrews' Extreme, Superstrong Backbones Are the Stuff of Legends," *Smithsonian Magazine*, April 30, 2020, https://www.smithsonianmag.com/smart-news/hero-shrews-extreme-superstrong-backbones-are-stuff-legend-180974778/.
7. Lents, *Human Errors*, 30.
8. Abby Hafer, *The Not-So-Intelligent Designer: Why Evolution Explains the Human Body and Intelligent Design Does Not* (Eugene, OR: Cascade Books, 2015), 179.
9. Lents, *Human Errors*, 30.
10. Alice Roberts, *The Incredible Unlikeliness of Being: Evolution and the Making of Us* (London: Quercus, 2014), 169.
11. Roberts, *The Incredible Unlikeliness of Being*, 169.

CHAPTER 6: MOUTH, NOSE, AND THROAT

1. S. C. Farina, E. A. Kane, and L. P. Hernandez, "Multifunctional Structures and Multistructural Functions: Integration in the Evolution of Biomechanical Systems," *Integrative and Comparative Biology* 59, no. 2 (August 2019): 340, https://doi.org/10.1093/icb/icz095.

2. Symposium S8: "Multifunctional Structures and Multistructural Functions: Functional Coupling and Integration in the Evolution of Biomechanical Systems," January 6, 2019, Society of Integrative & Comparative Biology, https://sicb.org /meetings/sicb-annual-meeting-2019/list-of-symposiums/symposium-s8/.

3. Stuart C. Burgess, "How Multifunctioning Joints Produce Highly Agile Limbs in Animals with Lessons for Robotics," *Biomimetics* 9, no. 9 (2024): 529, https://doi.org/10.3390/biomimetics9090529.

4. Minoru Hiran, "Vocal Mechanisms in Singing: Laryngological and Phoniatric Aspects," *Journal of Voice* 2, no. 1 (1988): 51–69, https://doi.org/10.1016/S0892 -1997(88)80058-4.

5. Andrea Rinaldi, "The Scent of Life: The Exquisite Complexity of the Sense of Smell in Animals and Humans," *EMBO Reports* 8, no. 7 (June 2007): 629–633, https://doi.org/10.1038/sj.embor.7401029.

6. Claire A. de March et al., "Engineered Odorant Receptors Illuminate the Basis of Odor Discrimination," *Nature* 635 (2024): 499–508, https://doi.org/10.1038 /s41586-024-08126-0.

7. Abby Hafer, *The Not-So-Intelligent Designer: Why Evolution Explains the Human Body and Intelligent Design Does Not* (Eugene, OR: Cascade Books, 2015), 72.

8. Interestingly, evolutionist Alice Roberts has conceded that deaths by choking are relatively rare and that choking is not that much different between humans and other animals. Alice Roberts, *The Incredible Unlikeliness of Being: Evolution and the Making of Us* (London: Quercus, 2014), 130.

9. Hafer, *The Not-So-Intelligent Designer*, 73.

10. For a more in-depth look at the problems facing neutral evolution, see Stephen C. Meyer, *Darwin's Doubt: The Explosive Origin of Animal Life and the Case for Intelligent Design* (New York: HarperOne, 2013), 321–329; and Michael J. Behe, *Darwin Devolves: The New Science About DNA That Challenges Evolution* (New York: HarperOne, 2019), 94–102.

Chapter 7: The Jaw

1. Sergio Soriano-Valero et al., "Systematic Review of Chewing Simulators: Reality and Reproducibility of *in vitro* Studies," *Journal of Clinical and Experimental Dentistry* 12, no. 12 (December 2020): e1189-e1195, doi:10.4317/jced.57279.

2. D. Raabe, A. Harrison, A. Ireland, K. Alemzadeh, J. Sandy, S. Dogramadzi, C. Melhuish, and S. Burgess, "Improved Single- and Multi-Contact Life-Time Testing of Dental Restorative Materials Using Key Characteristics of the Human Masticatory System and a Force/Position-Controlled Robotic Dental Wear Simulator," *Bioinspiration & Biomimetics* 7, no. 1 (2012): 1–17, doi: 10.1088/1748 -3182/7/1/016002.

3. Royal Society Summer Science Exhibition, London, 2009, https://www.imperial .ac.uk/events/113399/the-royal-society-summer-science-exhibition/.

4. Stephen Wroe et al., "The Craniomandibular Mechanics of Being Human," *Proceedings of the Royal Society B* 277 (June 23, 2010): 3579–3586, https://doi.org /10.1098/rspb.2010.0509.

5. J. W. Osborn, "Features of Human Jaw Design Which Maximize the Bite Force," *Journal of Biomechanics* 29, no. 5 (1996): 589–595, https://doi. org/10.1016/0021-9290(95)00117-4.

6. Elizabeth Steell, "Comparative Study of the Muscles of Mastication in Anthropoid Primates," *Journal of Morphology and Anatomy* 7, no. 2 (2023), https://www.hilarispublisher.com/open-access/comparative-study-of-the -muscles-of-mastication-in-anthropoid-primates-98134.html.

7. Neal Anthwal and Abigail S. Tucker, "The TMJ Disc Is a Common Ancestral Feature in All Mammals, as Evidenced by the Presence of a Rudimentary Disc During Monotreme Development," *Frontiers in Cell and Developmental Biology* 8 (May 2020): 356, https://doi.org/10.3389/fcell.2020.00356.

8. Abby Hafer, *The Not-So-Intelligent Designer: Why Evolution Explains the Human Body and Intelligent Design Does Not* (Eugene, OR: Cascade Books, 2015), 148.

9. It is well established that ancient populations experienced fewer instances of dental malocclusions. Some of this, as noted, is attributed to more robust jaw development due to diets involving less soft food. The increased grinding from the heartier diets also wore down some teeth, which is thought to have lessened the problem of tooth crowding. Genetic factors (microevolutionary changes) may also have played a role in the increase in malocclusion. For a recent analysis of a complex subject, see J. R. Herrera-Atoche, J. C. Chatters, and A. Cucina, "Unexpected Malocclusion in a 13,000-Year-old Late Pleistocene Young Woman from Mexico," *Scientific Reports* 12, article 3997 (2022), https://doi.org/10.1038/s41598-022-07941-7.

10. A. Evans et al., "A Simple Rule Governs the Evolution and Development of Hominin Tooth Size," *Nature* 530 (2016): 477–480, https://doi.org/10.1038 /nature16972.

Chapter 8: The Middle Ear

1. Stephen J. Elliott and Christopher A. Shera, "The Cochlea as a Smart Structure," *Smart Materials and Structures* 21, no. 6 (June 2012): 064001, doi: 10.1088/0964 -1726/21/6/064001.

2. Elliott and Shera, "The Cochlea as a Smart Structure."

3. C. J. Sumner et al., "Mammalian Behavior and Physiology Converge to Confirm Sharper Cochlear Tuning in Humans," *PNAS* 115, no. 44 (October 30, 2018): 11322–11326, https://doi.org/10.1073/pnas.1810766115.

4. Brian Cox, "Biology: Evolution of Hearing," *BBC National Curriculum KS3 / KS4*, accessed March 13, 2025, https://www.bbc.co.uk/teach/class-clips-video/articles /zmwjmfr.

5. Riley Black, "How to Make a Mammal in Nine Evolutionary Steps," *Smithsonian Magazine*, November 12, 2024, https://www.smithsonianmag.com/science -nature/how-to-make-a-mammal-in-nine-evolutionary-steps-180985289/ .

6. Margaux Schmeltz et al., "The Human Middle Ear in Motion: 3D Visualization and Quantification Using Dynamic Synchrotron-Based X-Ray Imaging," *Communications Biology* 7, article 157 (2024), https://doi.org/10.1038 /s42003-023-05738-6.

7. Andy McIntosh, "The Wonder of Hearing," accessed November 2025, https://apologeticsforum.org/mtgs18/WonderHearing.pdf.

Chapter 9: The Eye

1. M. Ponnavaillo and V. P. Kumar, "The Artificial Eye," *IEEE Potentials* 18, no. 5 (December 1999/January 2000): 33–35, doi: 10.1109/45.807278.

2. Hiromi Kobayashi and Shiro Kohshima, "Unique Morphology of the Human Eye and Its Adaptive Meaning: Comparative Studies on External Morphology of the Primate Eye," *Journal of Human Evolution* 40, no. 5 (June 2001): 419–435, doi: 10.1006/jhev.2001.0468.

3. For more on the exquisite and intricate design of the eye, see Michael Behe, *Darwin's Black Box* (New York: Free Press, 1996), 15–25 and 36–39; and Jonathan Wells, *Zombie Science: More Icons of Evolution* (Seattle, WA: Discovery Institute Press, 2017), chap. 7.

4. Charles Darwin, "Difficulties on Theory," in *On the Origin of Species by Means of Natural Selection, or the Preservation of Favoured Races in the Struggle for Life* (London: John Murray, 1859), chap. 6.

5. Behe, *Darwin's Black Box*, 22.

6. Richard Dawkins, *The Blind Watchmaker: Why the Evidence of Evolution Reveals a Universe Without Design* (New York: W. W. Norton, 1986), 93. Emphasis in original.

7. Richard Dawkins, *The Greatest Show on Earth* (London: Bantam Press, 2009), 353–354.

8. For more on evolutionists' bad-design claims, see Jonathan Wells, *Zombie Science*, 140–142.

9. Abby Hafer, *The Not-So-Intelligent Designer: Why Evolution Explains the Human Body and Intelligent Design Does Not* (Eugene, OR: Cascade Books, 2015), 111.

10. Wells, *Zombie Science*, 143–144.

11. Erez Ribak, "The Purpose of Our Eyes' Strange Wiring Is Unveiled," *Scientific American*, March 18, 2015, https://www.scientificamerican.com/article/the-purpose-of-our-eyes-strange-wiring-is-unveiled/. First published as Erez Ribak, "Look, Your Eyes Are Wired Backwards: Here's Why," *The Conversation*, March 13, 2015, https://theconversation.com/look-your-eyes-are-wired-backwards-heres-why-38319.

12. Erez Ribak, Amichai Labin, Shadi Safuri, and Ido Perlman, "Sorting of Colors in the Retina," presentation, American Physical Society meeting, San Antonio, TX, March 5, 2015, https://meetings.aps.org/Meeting/MAR15/Session/S47.2.

13. Amichai Labin and Erez Ribak, "Retinal Glial Cells Enhance Human Vision Acuity," *Physical Review Letters* 104 (April 16, 2010): 158102, https://doi.org/10.1103/PhysRevLett.104.158102.

14. Nathan H. Lents, *Human Errors: A Panorama of Our Glitches, from Pointless Bones to Broken Genes* (New York: Houghton Mifflin Harcourt, 2018), 5.

15. Hafer, *The Not-So-Intelligent Designer*, 70.

Chapter 10: Skin

1. Nina G. Jablonski, "Human Skin Pigmentation as an Example of Adaptive Evolution," *Proceedings of the American Philosophical Society* 156, no. 1 (March 2012): 45–57, https://www.jstor.org/stable/23558077 .

2. Monty Lyman, *The Remarkable Life of the Skin: An Intimate Journey Across Our Largest Organ* (New York: Grove Atlantic, 2019).

3. Lishi Li, et al. "The Functional Organization of Cutaneous Low-Threshold Mechanosensory Neurons," *Cell* 147, no. 7 (December 2011): 1615, doi: 10.1016/j.cell.2011.11.027.

4. Steven A. Brown et al., "Molecular Insights into Human Daily Behavior," *PNAS* 105, no. 5 (February 5, 2008): 1602–1607, https://doi.org/10.1073/pnas .0707772105.

5. Maksim V. Plikus et al., "The Circadian Clock in Skin: Implications for Adult Stem Cells, Tissue Regeneration, Cancer, Aging, and Immunity," *Journal of Biological Rhythms* 30, no. 3 (January 13, 2015), https://doi.org/10.1177 /0748730414563537.

6. P. D. Meglio, G. K. Perera, and F. O. Nestle, "The Multitasking Organ: Recent Insights into Skin Immune Function," *Immunity* 35 (2011): 857–869, doi: 10.1016/j.immuni.2011.12.003.

7. Alice Roberts, "Making the 'Perfect Body,'" *Science Museum Blog*, June 11, 2018, https://blog.sciencemuseum.org.uk/making-the-perfect-body/.

8. Abby Hafer, *The Not-So-Intelligent Designer: Why Evolution Explains the Human Body and Intelligent Design Does Not* (Eugene, OR: Cascade Books, 2015), 179.

Chapter 11: Birth Biomechanics

1. James A. Ashton-Miller and John O. Delancey, "On the Biomechanics of Vaginal Birth and Common Sequelae," *Annual Review of Biomedical Engineering* 11 (2009): 163.

2. Michele J. Grimm, "Forces Involved with Labor and Delivery—A Biomechanical Perspective," *Annals of Biomedical Engineering* 49 (2021): 1819–1835, https://link .springer.com/article/10.1007/s10439-020-02718-3.

3. Diego R. Higueras-Ruiz et al., "What Is an Artificial Muscle? A Comparison of Soft Actuators to Biological Muscles," *Bioinspiration and Biomimetics* 17, no. 1 (December 23, 2021), doi: 10.1088/1748-3190/ac3adf.

4. Oliver Ami et al., "Three-Dimensional Magnetic Resonance Imaging of Fetal Head Molding and Brain Shape Changes During the Second Stage of Labor," *PLoS One* 14, no. 5 (May 2019): e0215721, https://doi.org/10.1371/journal .pone.0215721.

5. Sana T Saiyed et al., "Stillbirth Rates Across Three Ape Species in Accredited American Zoos," *American Journal of Primatology* 80, no. 6 (June 2018), doi: 10.1002/ajp.22870.

6. For those interested in delving into this theological question of why there is pain and death in a world created by a good and all-powerful God, a good place to begin is C. S. Lewis's short work *The Problem of Pain* (1940). There are many other well-thought-out treatments of the question. An early treatment can be found in Augustine's *City of God* (circa AD 413–426). And a particularly in-depth more recent treatment is Eleonore Stump's *Wandering in Darkness: Narrative and the Problem of Suffering* (Oxford, UK: Oxford University Press, 2010).

7. Abby Hafer, *The Not-So-Intelligent Designer: Why Evolution Explains the Human Body and Intelligent Design Does Not* (Eugene, OR: Cascade Books, 2015), 47. For a similar argument, see Alice Roberts, "Can Science Make Me Perfect?," *BBC4*, June 13, 2018, audio, 1:29:00, https://www.bbc.co.uk/programmes/b0b6q3qy.

Chapter 12: The Blood Circulatory System

1. Sian E. Harding, "In Search of the Impossible Machine, the Artificial Heart," *The MIT Press Reader*, https://thereader.mitpress.mit.edu/in-search-of-the-impossible

-machine-the-artificial-heart/. The article is excerpted from Harding's book, *The Exquisite Machine: The New Science of the Heart* (Cambridge, MA: The MIT Press, 2022), 141–147. Ian Peate, editor-in-chief of the *British Journal of Nursing*, strikes a similar chord in "The Heart: An Amazing Organ," *British Journal of Healthcare Assistants* 15, no. 2 (March 9, 2021), https://doi.org/10.12968/bjha.2021.15.2.72.

2. R. Monahan-Earley, A. M. Dvorak, and W. C. Aird, "Evolutionary Origins of the Blood Vascular System and Endothelium," *Journal of Thrombosis and Haemostasis* 11, no. 1 (June 30, 2013): 46–66, https://onlinelibrary.wiley.com/doi/10.1111/jth.12253.

3. Casey Luskin, "The Vertebrate Animal Heart: Unevolvable, Whether Primitive or Complex," *IDEA Center*, 2004, http://www.ideacenter.org/contentmgr /showdetails.php/id/1113.

4. Richard Dawkins, *The Greatest Show on Earth: The Evidence for Evolution* (New York: Free Press, 2009), 371.

5. Alice Roberts, "Can Science Make Me Perfect?," *BBC4*, June 13, 2018, audio, 1:29:00, https://www.bbc.co.uk/programmes/b0b6q3qy.

CHAPTER 13: THE DIGESTIVE SYSTEM

1. Ilkay Sensoy, "A Review on the Food Digestion in the Digestive Tract and the Used *in vitro* Models," *Current Research in Food Science* 4, no. 2 (April 2021): 308–319.

2. "The Developing Gut Does the Twist," *Nature* 561 (September 6, 2018): 8. The news article reports on the research findings of Aravind Sivakumar et al., "Midgut Laterality Is Driven by Hyaluronan on the Right," *Developmental Cell* 46, no. 5 (September 10, 2018): 533–551.e5, https://doi.org/10.1016/j.devcel.2018.08.002.

3. Abby Hafer, *The Not-So-Intelligent Designer: Why Evolution Explains the Human Body and Intelligent Design Does Not* (Eugene, OR: Cascade Books, 2015), 178.

4. Charles Darwin, *The Descent of Man, and Selection in Relation to Sex* (London: John Murray, 1871), 27.

5. Richard J. A. Berry, "The True Caecal Apex, or the Vermiform Appendix: Its Minute and Comparative Anatomy," *Journal of Anatomy and Physiology* 35 (1900): 83–100, https://pmc.ncbi.nlm.nih.gov/articles/PMC1287282/. "The vermiform appendix of Man is not, therefore, a vestigial structure," Berry concludes. "On the contrary, it is a specialized part of the alimentary canal" (98).

6. Jonathan Wells cites Berry and several other more recent scientific papers exploring the dual functions of the appendix, in *Zombie Science: More Icons of Evolution* (Seattle, WA: Discovery Institute Press, 2017), 116–119.

7. R. Randal Bollinger et al., "Biofilms in the Large Bowel Suggest an Apparent Function of the Human Vermiform Appendix," *Journal of Theoretical Biology* 249, no. 4 (December 21, 2007): 826–831.

8. Jan-Olaf Gebbers and Jean-Albert Laissue, "Bacterial Translocation in the Normal Human Appendix Parallels the Development of the Local Immune System," *Annals of the New York Academy of Sciences* 1029, no. 1 (January 12, 2006), https://doi.org/10.1196/annals.1309.015.

9. Loren G. Martin, "What Is the Function of the Human Appendix? Did It Once Have a Purpose That Has Since Been Lost?" *Scientific American* (October 21, 1999), https://www.scientificamerican.com/article/what-is-the-function-of -the-human-appendix-did-it-once-have-a-purpose-that-has-since-been-lost/.

10. Tarequl Islam et al., "Exploring the Immunological Role of the Microbial Composition of the Appendix and the Associated Risks of Appendectomies," *Journal of Personalized Medicine* 15, no. 3 (March 13, 2025), https://doi.org/10.3390/jpm15030112.

11. For more on the subject see Wells, *Zombie Science*, chap. 6.

CHAPTER 14: MUSCLES AND TENDONS

1. George Székely, "A Perfect Design: The Multifunctional Muscle," *Behavioral and Brain Sciences* 12, no. 4 (1989): 668–669.

2. "Tendon," *Britannica*, accessed July 8, 2025, https://www.britannica.com/science/tendon.

3. Oluwaseun A. Araromi and Stuart C. Burgess, "A Finite Element Approach for Modelling Multilayer Unimorph Dielectric Elastomer Actuators with Inhomogeneous Layer Geometry," *Smart Materials and Structures* 21, no. 3 (February 10, 2012).

4. Nathan H. Lents, *Human Errors: A Panorama of Our Glitches, from Pointless Bones to Broken Genes* (New York: Houghton Mifflin Harcourt, 2018), 25.

5. Yoann Demangeot et al., "The Load Borne by the Achilles Tendon During Exercise: A Systematic Review of Normative Values," *Scandinavian Journal of Medicine & Sports* 33, no. 2 (February 2023): 110–126, https://doi.org/10.1111/sms.70140.

CHAPTER 15: THE NERVOUS SYSTEM

1. Peter Sterling and Simon Laughlin, *Principles of Neural Design* (Cambridge, MA: MIT Press, 2015), 433, https://archive.org/details/principlesofneur0000ster/page/n5/mode/2up.

2. Sterling and Laughlin, *Principles of Neural Design*, 440.

3. Sterling and Laughlin, *Principles of Neural Design*, 1.

4. Selmer Bringsjord, Paul Bello, and David Ferrucci, "Creativity, the Turing Test, and the (Better) Lovelace Test," in *The Turing Test: The Elusive Standard of Artificial Intelligence*, ed. James H. Moor (Boston: Kluwer Academic Publishers, 2003), 215–239.

5. Robert J. Marks, *Non-Computable You: What You Do That Artificial Intelligence Never Will* (Seattle, WA: Discovery Institute Press, 2022), 42–43, 56.

6. Erik J. Larson, "Just When Human Reason Is Most Productive—AI Makes Things Up," *Mind Matters*, June 18, 2025, https://mindmatters.ai/2025/06/just-when-human-reason-is-most-productive-ai-makes-things-up/.

7. "Superintelligent AI Is Still a Myth," *Mind Matters*, February 5, 2020, https://mindmatters.ai/2020/02/superintelligent-ai-is-still-a-myth/.

8. Bente Pakkenberg et al., "Aging and the Human Neocortex," *Experimental Gerontology* 38, no. 1–2 (January 2003): 95–99, https://doi.org/10.1016/S0531-5565(02)00151-1.

9. Nathan H. Lents, *Human Errors: A Panorama of Our Glitches, from Pointless Bones to Broken Genes* (New York: Houghton Mifflin Harcourt, 2018), 13–16.

10. Jerry A. Coyne, *Why Evolution Is True* (Oxford, UK: Oxford University Press, 2009), 87, 88.

11. Richard Dawkins, *The Greatest Show on Earth: The Evidence for Evolution* (New York: Free Press, 2009), 364.

CHAPTER 16: BIOLOGICAL SYSTEMS CANNOT SELF-ORGANIZE

1. Theodore von Kármán, quoted in "Theodore von Kármán: National Medal of Science—Engineering," National Science & Technology Medals Foundation, 2025, https://nationalmedals.org/laureate/theodore-von-karman/.
2. S. C. Burgess, "A Study of the Efficiency of a Double-Action Worm Gear Set," *Proceedings of the Institution of Mechanical Engineers, Part G: Journal of Aerospace Engineering* 206, no. 2 (July 1992): 81–91, https://doi.org/10.1243/PIME _PROC_1992_206_245_02.
3. Alexander Graham Bell is conventionally credited with inventing the telephone, but the Library of Congress reports, "Antonio Meucci, an Italian immigrant, began developing the design of a talking telegraph or telephone in 1849. In 1871, he filed a caveat (an announcement of an invention, often a precursor to a patent) for his design of a talking telegraph. Due to financial hardships, Meucci could not renew his caveat. His role in the invention of the telephone was overlooked until the United States House of Representatives passed a Resolution on June 11, 2002, honoring Meucci's contributions and work. You can read the resolution (107th Congress, H Res 269) on Congress.gov." See "Who Is Credited with Inventing the Telephone," Library of Congress, https://www.loc.gov/everyday-mysteries /technology/item/who-is-credited-with-inventing-the-telephone/.

CHAPTER 17: EVOLUTIONARY THEORY IN DECLINE

1. Paul Davies, *The Fifth Miracle: The Search for the Origin and Meaning of Life* (New York: Simon and Schuster, 1998), xvii.
2. Stephen Jay Gould, "Evolution's Erratic Pace," *Natural History* 86, no. 5 (May 1977): 12–16.
3. Graham Lawton, "Why Darwin Was Wrong About the Tree of Life," *New Scientist* (January 21, 2009), https://www.newscientist.com/article/mg20126921 -600-why-darwin-was-wrong-about-the-tree-of-life/.
4. Günter Theißen, "Saltational Evolution: Hopeful Monsters Are Here to Stay," *Theory in Biosciences* 128, no. 1 (March 2009): 43–44. Internal references removed.
5. Lynn Margulis and Dorion Sagan, *Acquiring Genomes: A Theory of the Origins of the Species*, (New York: Basic Books, 2003), 29.
6. Richard Lewontin, "Billions and Billions of Demons," review of *The Demon-Haunted World: Science as a Candle in the Dark*, by Carl Sagan, *New York Review of Books*, January 9, 1997, 31. Emphasis in original.
7. Julian Huxley, *Evolution: The Modern Synthesis* (London: George Allen & Unwin Ltd, 1942), 457.
8. Charles Darwin to J. D. Hooker, February 1, 1871, *Darwin Correspondence Project*, Letter no. 7471, University of Cambridge, https://www.darwinproject.ac.uk /letter/DCP-LETT-7471.xml.
9. Alfred Russel Wallace, letter to the editor, *Light*, July 26, 1890, 462, available at Charles H. Smith, Alfred Russel Wallace Page, Western Kentucky University, https://people.wku.edu/charles.smith/wallace/S425A.htm.

10. Lord Kelvin, letter to the editor in response to "Lord Kelvin on Religion and Science," *The Times*, May 4, 1903, https://zapatopi.net/kelvin/papers/science_affirms_creative_power.html.

11. For more on the challenge the second law of thermodynamics poses for abiogenesis, see Charles B. Thaxton, Walter L. Bradley, and Roger L. Olsen, *The Mystery of Life's Origin: The Continuing Controversy* (Seattle: Discovery Institute Press, 2020), chaps. 7 and 8, as well as the supplementary chap. 14 by Brian Miller, "Thermodynamic Challenges to the Origin of Life." The book is a revised and expanded second edition of Thaxton, Bradley, and Olsen's seminal work *The Mystery of Life's Origin: Assessing Current Theories* (New York: Philosophical Library, 1984). See also Granville Sewell, *In the Beginning and Other Essays on Intelligent Design* (Seattle: Discovery Institute Press, 2010), 53–55, 63–81, 85–90.

12. Francis Crick, *Life Itself* (New York: Simon & Schuster, 1981), 88.

13. Paul Davies, "Life Force," *New Scientist* 163 (September 1999): 28.

14. Ker Than, "Smallest Genome of Living Creature Discovered," *Live Science*, October 14, 2022, https://www.livescience.com/1091-smallest-genome-living-creature-discovered.html.

15. For more on the information argument to design in the origin of life, see Stephen C. Meyer, *Signature in the Cell: DNA and the Evidence for Intelligent Design* (New York: HarperOne, 2009).

16. John I. Glass et al., "Minimal Cells—Real and Imagined," *Cold Spring Harbor Perspectives in Biology* (2017), https://cshperspectives.cshlp.org/content/early/2017/03/27/cshperspect.a023861.

17. Eric Anderson, "A Factory That Builds Factories That Builds Factories That…," *Evolution and Intelligent Design in a Nutshell* (Seattle: Discovery Institute Press, 2020), 65–86.

18. Brian Cox in video at "KS3 / KS4 Biology: The "Origins of Life on Earth," *BBC Bitesize*, https://www.bbc.co.uk/teach/class-clips-video/articles/zh8fcqt.

19. Daniel P. Glavin et al., "Abundant Ammonia and Nitrogen-Rich Soluble Organic Matter in Samples from Asteroid (101955) Bennu," *Nature Astronomy* (January 2025), https://www.nature.com/articles/s41550-024-02472-9.

20. Stefan Milovanovic, "Human DNA Detected in 2 Billion Year Old Meteorite," *Yahoo! News*, October 30, 2025, https://www.yahoo.com/news/articles/human-dna-detected-2-billion-211520610.html.

21. Jonathan Wells, *Zombie Science: More Icons of Evolution* (Seattle, WA: Discovery Institute Press, 2017), 67–69.

22. Wells, *Zombie Science*, 68–69.

23. See Michael J. Behe, *Darwin Devolves: The New Science About DNA That Challenges Evolution* (New York: HarperOne, 2019). There he argues that in cases where some improvement arises via a genetic mutation (e.g., antibiotic resistance), the mutation inevitably involves evolutionary degradation. The advantage is a localized ecological advantage gained by blunting or breaking something, and typically it comes at the cost of ecological flexibility/overall fitness.

24. Wells, *Zombie Science*, chap. 5, "Walking Whales."

25. Sean B. Carroll, "The Big Picture," *Nature* 409 (February 8, 2001): 669, https://doi.org/10.1038/35055637.

26. Scott Gilbert and Graham Budd are quoted in John Whitfield, "Biological Theory: Postmodern Evolution?," *Nature* 455 (September 17, 2008): 282, https://doi.org/10.1038/455281a.

27. Douglas H. Erwin and James W. Valentine, "'Hopeful Monsters,' Transposons, and Metazoan Radiation," *PNAS* 81, no. 17 (September 1984): 5482–5483.

28. James Valentine and Douglas Erwin, "Interpreting Great Developmental Experiments: The Fossil Record" in *Development as an Evolutionary Process*, eds. Rudolf A. Raff and Elizabeth C. Raff (New York: Alan R. Liss, 1985), 95.

29. Scott Gilbert, John Opitz, and Rudolf Raff, "Resynthesizing Evolutionary and Developmental Biology," *Developmental Biology* 173 (1996): 361.

30. Günter Bechly, "Fossil Friday: Discontinuities in the Fossil Record—A Problem for Neo-Darwinism," *Science & Culture Today*, May 10, 2024, https://scienceandculture.com/2024/05/fossil-friday-discontinuities-in-the-fossil-record-a-problem-for-neo-darwinism/.

31. The reality behind the peppered moth story is actually even less favorable to evolutionary theory than this. See Jonathan Wells, *Zombie Science*, 63–66.

32. Steve Jones, *Darwin's Ghost:* The Origin of Species *Updated* (New York: Random House, 2000), 98.

33. Alice Roberts, *The Incredible Unlikeliness of Being: Evolution and the Making of Us* (London: Heron Books, 2014), 171.

34. Nathan Lents, *Human Errors: A Panorama of Our Glitches, from Pointless Bones to Broken Genes* (London: Weidenfeld & Nicolson, 2008), 27.

35. Jeremy DeSilva, "Starting Off on the Wrong Foot: How Our Ape Ancestry Predisposes Us to Foot and Ankle Maladies," (lecture, AAAS annual meeting, Boston, MA, February 14–18, 2013). According to DeSilva, "The evolution of a stable [lever] structure from a grasping one has left us particularly susceptible to a variety of foot and ankle injuries."

36. For example, see David C. Marciano, Omid Y. Karkouti, and Timothy Palzkill, "A Fitness Cost Associated with the Antibiotic Resistance Enzyme SME-1 β-Lactamase," *Genetics* 176, no. 4 (August 2007): 2381–2392, https://doi.org/10.1534/genetics.106.069443; Ming An Shi et al., "Acetylcholinesterase Alterations Reveal the Fitness Cost of Mutations Conferring Insecticide Resistance," *BMC Evolutionary Biology* 4, no. 5 (February 6, 2004), https://doi.org/10.1186/1471-2148-4-5; Linus Sandegren et al., "Nitrofurantoin Resistance Mechanism and Fitness Cost in *Escherichia coli*," *Journal of Antimicrobial Chemotherapy* 62, no. 3 (September 2008): 495–503, https://doi.org/10.1093/jac/dkn222; Dan I. Andersson, "The Biological Cost of Mutational Antibiotic Resistance: Any Practical Conclusions?," *Current Opinion in Microbiology* 9, no. 5 (October 2006): 461–465.

37. Michael Behe, *Darwin Devolves*, 15–16.

38. *Darwin Devolves: The New Science About DNA That Challenges Evolution* (New York: HarperOne, 2019), vii.

39. Dustin J. Van Hofwegen, Carolyn J. Hovde, and Scott A. Minnich, "Rapid Evolution of Citrate Utilization by *Escherichia coli* by Direct Selection Requires citT and dctA," *Journal of Bacteriology*, 198, no. 7 (April 1, 2016): 1022–1034, https://doi.org/10.1128/jb.00831-15.

40. Behe assesses the results of Richard Lenski's now-famous long-term evolutionary experiment and deflates the hype surrounding it. See his books *Darwin Devolves*, 188–190 and *A Mousetrap for Darwin: Michael J. Behe Answers His Critics* (Seattle: Discovery Institute Press, 2020), 263–291, 394–410.

41. Behe, *Darwin Devolves*, 175. The genetics expert John Sanford argues something similar. At Cornell, Sanford was involved in developing the Biolistic Particle Delivery System, popularly known as the "gene gun." He also co-invented the pathogen-derived resistance (PDR) process and the genetic vaccination process. In the course of his scientific career he came to doubt evolutionary theory and wrote the book *Genetic Entropy* (2014) where he shows that information in DNA is gradually degrading from one generation to the next due to mutations, and that natural selection, far from leading to inevitable biological improvement, is not enough to stop the accumulation of harmful mutations. Yes, natural selection and life's ingenious error-correcting features slow down genetic degradation from one generation to the next, but they aren't enough to stop it altogether. Examples of genetic degradation in the human body include the genetic disorder cystic fibrosis. Individuals who suffer from these genetic disorders are, of course, no less human than anyone else, but none of us should mistake such genetic disorders as genetic mutations on the way to some wonderful new biological form, such as we routinely encounter in science fiction stories (e.g., the Incredible Hulk, X-Men) where genetic mutations lend the heroes special powers.

42. Design theorists are under no such psychological pressure. Since they see the manifold variety of life as the product of a masterful designing intelligence, and one who is free to work through either primary or secondary causes, they have no compulsion one way or the other when investigating cases of microevolution. Maybe the change is a function of the organism's preprogrammed capacity for variation within some preset range. Maybe the micromutation was directly designed? Or maybe it was a random genetic mutation, a devolutionary break that proved fortuitous in a specialized context. The design theorist is free simply to investigate and follow the evidence.

43. Behe, *The Edge of Evolution: The Search for the Limits of Darwinism* (New York: Free Press, 2007), 17–63, 74–77.

44. Jerry Coyne, "The Great Mutator," *The New Republic*, June 18, 2007, 38–44.

45. Aniket V. Gore et al., "An Epigenetic Mechanism for Cavefish Eye Degeneration," *Nature Ecology & Evolution* 2 (2018): 1155–1160, https://doi.org/10.1038/s41559-018-0569-4.

46. Charles Darwin to J. D. Hooker, July 22, 1879, *Darwin Correspondence Project*, Letter no. 12167, University of Cambridge, https://www.darwinproject.ac.uk/letter/?docId=letters/DCP-LETT-12167.xml. See also Richard J. A. Buggs, "The Deepening of Darwin's Abominable Mystery," *Nature Ecology & Evolution* 1, article 0169 (2017), https://doi.org/10.1038/s41559-017-0169.

47. David Grimaldi and Michael S. Engel, *Evolution of the Insects* (Cambridge, UK: Cambridge University Press, 2005), 302; Stanley A. Rice, *Encyclopedia of Evolution* (New York: Checkmark, 2007), 70; Stefanie De Bodt, Steven Maere, and Yves Van de Peer, "Genome Duplication and the Origin of Angiosperms," *Trends in Ecology and Evolution* 20 (2005): 591–597.

48. Günter Bechly, "Fossil Friday: The Carboniferous Explosion of Winged Insects," *Science & Culture Today*, August 9, 2024, https://scienceandculture.com/2024/08/fossil-friday-the-carboniferous-explosion-of-winged-insects/.

49. A. T. Conn, S. C. Burgess, and C. S. Ling, "Design of a Parallel Crank-Rocker Flapping Mechanism for Insect-Inspired Micro Air Vehicles," *Proceedings of the Institution of Mechanical Engineers, Part C: Journal of Mechanical Engineering Science* 221, no. 10 (October 1, 2007): 1211–1222, https://doi.org/10.1243/09544062JMES517.

50. Günter Bechly, "Fossil Friday: The Abrupt Origin of Winged Insects," *Science & Culture Today,* March 24, 2023, https://scienceandculture.com/2023/03/fossil-friday-the-abrupt-origin-of-winged-insects/.

51. David E. Alexander, "A Century and a Half of Research on the Evolution of Insect Flight," *Arthropod Structure & Development* 47, no. 4 (July 2018): 322–327, https://doi.org/10.1016/j.asd.2017.11.007.

52. Richard Dawkins, *The Blind Watchmaker* (London: Penguin, 1988), x.

CHAPTER 18: INTELLIGENT DESIGN ASCENDING

1. Leonardo da Vinci, *The Complete Notebooks of Leonardo Da Vinci*, Vol. 2, Book XIV, "Anatomy, Zoology and Physiology," trans. Jean Paul Richter (1888), 837, https://www.gutenberg.org/cache/epub/4999/pg4999-images.html.

2. "Peer-Reviewed Articles Supporting Intelligent Design," Discovery Institute Center for Science and Culture, https://www.discovery.org/id/peer-review/, accessed June 18, 2025.

3. Stuart Burgess, "Universal Optimal Design in the Vertebrate Limb Pattern and Lessons for Bioinspired Design," *Bioinspiration and Biomimetics* 19, no. 5 (September 2024), doi 10.1088/1748-3190/ad66a3.

4. American Museum of Natural History, "Whales: Giants of the Deep—Breathing, Feeding, & Moving in Water," 2013, pdf available at https://share.google/u1nGOiEVEgb2opG4i; "Movement," Museum of New Zealand, accessed October 2, 2025, https://www.tepapa.govt.nz/about/touring-exhibitions/whales-tohora/whale-lab/movement.

5. Burgess, "Universal Optimal Design in the Vertebrate Limb Pattern," 24.

6. Charles Darwin, *On the Origin of Species by Means of Natural Selection, or the Preservation of Favoured Races in the Struggle for Life* (London: John Murray, 1859), 189. Darwin deserves credit for offering a means of testing and potentially falsifying his theory, but notice that he has shifted the burden of proof off of his theory and onto anyone opposing it, as if his theory should be considered "innocent until proven guilty."

7. Michael Behe, *Darwin's Black Box* (New York: Free Press, 1996).

8. Behe criticized the co-option scenario ("the principle of tinkering") in *Darwin's Black Box*, 40, and in *Darwin Devolves: The New Science About DNA That Challenges Evolution* (New York: HarperOne, 2019), 81, 89, 100. See also his *A Mousetrap for Darwin: Michael J. Behe Answers His Critics* (Seattle, WA: Discovery Institute Press, 2020), 150–152, 300, 396–398, 447, 486–487.

9. For a summary of the problems with the co-option explanation for the origin of the bacterial flagellum, see Stephen Dilley, Casey Luskin, Brian Miller, and Emily Reeves, "On the Relationship Between Design and Evolution," *Religions* 14, no. 7 (2023): 850, https://doi.org/10.3390/rel14070850.

10. If Behe's argument against evolutionary theory and for intelligent design is true then, of course, it is not falsifiable *in fact* but only *in principle*—just as, for instance,

the hypothesis of plate tectonics, being true, is not falsifiable in fact but only in principle. This is hardly a mark against its status as a properly scientific argument.

11. Marcus Tullius Cicero, *The Nature of the Gods* [45 BC], trans. P. G. Walsh (New York: Oxford University Press, 1998), 78, https://archive.org/details /natureofgods0000cice/page/78/mode/2up?q=clock.

12. William Paley, *Natural Theology: Or Evidences of the Existence and Attributes of the Deity, Collected from the Appearances of Nature* [1802], eds. Matthew D. Eddy and David Knight (Oxford, UK: Oxford University Press, 2008). Paley writes, "Every indication of contrivance, every manifestation of design which existed in the watch, exists in the work of nature, with the difference on the side of nature of being greater and more, and that in a degree which exceeds all computation," 16, https://archive.org/details/B-001-001-222/page/n55/mode/2up?q=atheism.

13. Behe, *Darwin's Black Box*.

14. Stuart Burgess, "A Review of Linkage Mechanisms in Animal Joints and Related Bioinspired Designs," *Bioinspiration and Biomimetics* 16, no. 4 (June 2021); Stuart C. Burgess, "How Multifunctioning Joints Produce Highly Agile Limbs in Animals with Lessons for Robotics," *Biomimetics* 9, no. 9 (2024): 529, https://doi .org/10.3390/biomimetics9090529; S. Burgess at al., "A Bio-Inspired Arched Foot with Individual Toe Joints and Plantar Fascia," *Biomimetics* 8 (2023): 455.

15. Steve Laufmann and Howard Glicksman, *Your Designed Body* (Seattle, WA: Discovery Institute Press, 2022).

16. Burgess, "How Multifunctioning Joints Produce Highly Agile Limbs," 529. Internal references removed.

17. Stephen C. Meyer, *Signature in the Cell* (New York: HarperOne, 2010).

18. Andy C. McIntosh, "Information and Entropy—Top-down or Bottom-up Development in Living Systems," *International Journal of Design & Nature and Ecodynamics* 4, no. 4 (2009): 351–385; A. C. McIntosh, "Information and Thermodynamics in Living Systems," in *Biological Information—New Perspectives: Proceedings of a Symposium Held May 31, 2011 through June 3, 2011 at Cornell University* (Hackensack, NJ: World Scientific, 2013), 179–201.

19. Granville Sewell, *In the Beginning and Other Essays on Intelligent Design* (Seattle, WA: Discovery Institute Press, 2010), 53–55, 63–81, 85–90.

20. K. Laland et al., "Does Evolutionary Theory Need a Rethink?," *Nature* 514 (October 2014):161–164.

21. John Maynard Smith, "Theories of Sexual Selection," *Trends in Ecology and Evolution* 6 (May 1991):146–151.

22. Charles Darwin, *The Origin of Species* [1859] (New York: P. F. Collier and Son, 1937), 199–200, https://archive.org/details/in.ernet.dli.2015.206064/page /n205/mode/2up?q=%22delight+man%22.

23. Charles Darwin to Asa Gray, April 3, 1860, *Darwin Correspondence Project*, Letter no. 2743, University of Cambridge, https://www.darwinproject.ac.uk /letter?docId=letters/DCP-LETT-2743.xml.

24. Charles Darwin, *The Descent of Man, and Selection in Relation to Sex* [1871] (London: John Murray, 1901), 608, https://archive.org/details/ncbs.BB-001 _0_0_0_1/page/n625/mode/2up?q=%22exquisite+patterns%22.

25. Darwin, *The Descent of Man*, 609.

26. William Homan Thorpe, *Bird-Song: The Biology of Vocal Communication and Expression in Birds* (Cambridge, UK: Cambridge University Press, 1961), 64, 63, https://archive.org/details/birdsongbiologyo0000whth_p8u3/page/64/mode /2up?q=%22extreme+purity%22.

27. Stuart Burgess, *Hallmarks of Design* (United Kingdom: DayOne, 2000).

28. C. K. Catchpole and P. J. B. Slater, *Bird Song: Biological Themes and Variations* (Cambridge, UK: Cambridge University Press, 1995), 11.

29. E. A. Armstrong, *A Study of Bird Song* (Oxford University Press, 1963), 30, 35.

30. W. H. Thorpe et al., "Duetting and Antiphonal Song in Birds: Its Extent and Significance," *Behaviour* Supplement 18 (1972): 148.

31. Catchpole and Slater, *Bird Song*, 116.

32. Steve Jones, *Darwin's Ghost:* The Origin of Species *Updated* (New York: Random House, 2000), 98.

33. Alice Roberts, "The Origins of Us: Human Anatomy and Evolution" (lecture, University of Birmingham, February 10, 2012), https://www.birmingham .ac.uk/accessibility/transcripts/origins-alice-roberts.

34. Steven Pinker, *The Language Instinct* (New York: Harper Perennial, 1994), 378.

35. Alfred Russel Wallace, *The World of Life: A Manifestation of the Creative Power, Directive Mind, and Ultimate Purpose* (London: Chapman and Hall, 1914), 316.

36. Stephen Jay Gould and Richard C. Lewontin, "The Spandrels of San Marco and the Panglossian Paradigm: A Critique of the Adaptationist Programme," *Proceedings of the Royal Society of London B* 205 (September 21, 1979): 581–598, https://doi.org/10.1098/rspb.1979.0086.

CHAPTER 19: ULTIMATE ENGINEERING THROUGHOUT NATURE

1. Guruaj Fattepur, et al., "Bio-Inspired Designs: Leveraging Biological Brilliance in Mechanical Engineering—An Overview," *3 Biotech* 14, article no. 312 (November 2024), https://doi.org/10.1007/s13205-024-04153-w.

2. Ionkov Latchesar and Bradley Settlemyer, "DNA: The Ultimate Data-Storage Solution," *Scientific American* (May 28, 2021).

3. Nan Zhang et al., "Toward Rational Algorithmic Design of Collagen-Based Biomaterials Through Multiscale Computational Modelling," *Current Opinion in Chemical Engineering* 24 (June 2019): 79–87, https://doi.org/10.1016/j.coche .2019.02.011. Internal references removed.

4 S. C. Burgess, "The Ranking of Efficiency of Structural Layouts Using Form Factors, Part I: Design for Stiffness," *Proceedings of the Institution of Mechanical Engineers, Part C: Journal of Mechanical Engineering Science* 212, no. 2 (1998): 117–128, https://doi.org/10.1243/0954406981521088.

5 Stuart Burgess, "Universal Optimal Design in the Vertebrate Limb Pattern and Lessons for Bioinspired Design," *Bioinspiration and Biomimetics* 19, no. 5 (September 2024), doi 10.1088/1748-3190/ad66a3.

6. S. C. Burgess, "The Ranking of Efficiency of Structural Layouts," 117–128.

7. Stuart Burgess, "Universal Optimal Design in the Vertebrate Limb Pattern."

8. Stuart Burgess, "A Review of Linkage Mechanisms in Animal Joints and Related Bioinspired Designs," *Bioinspiration and Biomimetics* 16, no. 4 (June 2021), doi: 10.1088/1748-3190/abf744.

9. Bottlenose dolphins manage higher click rates, but the specialized system only works in water.

10. Atticus Pinzon-Rodriguez, Staffan Bensch, and Rachel Muheim, "Expression Patterns of Cryptochrome Genes in Avian Retina Suggest Involvement of Cry4 in Light-Dependent Magnetoreception," *Journal of the Royal Society Interface* 15 (March 28, 2018), http://doi.org/10.1098/rsif.2018.0058.

11. Chris P. F. Redfern and Richard M. Bevan, "Overland Movement and Migration Phenology in Relation to Breeding of Arctic Terns *Sterna paradisaea*," *Ibis* 162, no. 2 (April 2020): 373–380, https://doi.org/10.1111/ibi.12723.

12. Jason Bitel, "Monarch Butterflies Migrate 3,000 Miles—Here's How," *National Geographic*, October 16, 2017, https://www.nationalgeographic.com/animals/article/monarch-butterfly-migration.

13. For more on the wonders of animal migration and their design implications, see Eric Cassell, *Animal Algorithms: Evolution and the Mysterious Origin of Ingenious Instincts* (Seattle, WA: Discovery Institute Press, 2021), chaps. 2 and 3.

14. Farshid Guilak, "The Slippery Slope of Arthritis," *Arthritis & Rheumatology* 52, no. 6 (June 2005): 1632–1633.

15. Ola Hössjer, Günter Bechly, and Ann Gauger, "On the Waiting Time Until Coordinated Mutations Get Fixed in Regulatory Sequences," *Journal of Theoretical Biology* 524 (September 7, 2021), https://doi.org/10.1016/j.jtbi.2021.110657.

16. Steve Jones, *Darwin's Ghost* (New York: Random House, 1999), 70.

17. Jobran Chebib et al., "An Estimate of Fitness Reduction from Mutation Accumulation in a Mammal Allows Assessment of the Consequences of Relaxed Selection," *PLoS Biology* 22, no. 9 (September 26, 2024), https://doi.org/10.1371/journal.pbio.3002795.

CHAPTER 20: ULTIMATE DIVERSITY THROUGHOUT NATURE

1. G. V. Mazzetta, P. Christiansen, and R. A. Fariña, "Giants and Bizarres: Body Size of Some Southern South American Cretaceous Dinosaurs," *Historical Biology* 16 (May 17, 2004): 71–83, https://doi.org/10.1080/08912960410001715132.

2. Stuart C. Burgess et al., "A Study of Mechanical Configuration Optimisation in Micro-Systems," *Journal of Research in Engineering Design*, 9 (1997): 46–60, https://doi.org/10.1007/BF01607057.

3. J. A. Goldbogen et al., "Extreme Bradycardia and Tachycardia in the World's Largest Animal," *PNAS* 116, no. 50 (November 25, 2019): 25329–25332, https://doi.org/10.1073/pnas.1914273116.

4. Q. Zhong et al., "Tunable Stiffness Enables Fast and Efficient Swimming in Fish-Like Robots," *Science Robotics* 6 (August 2021), doi: 10.1126/scirobotics.abe4088.

5. R. J. Lock, S. C. Burgess, and R. Vaidyanathan, "Multi-Modal Locomotion: From Animal to Application," *Bioinspiration and Biomimetics* 9, no. 1 (2014), doi: 10.1088/1748-3182/9/1/011001.

6. Casper J. van der Kooi et al., "Functional Optics of Glossy Buttercup Flowers," *Journal of the Royal Society Interface* 14, no. 127 (February 2017), https://doi.org/10.1098/rsif.2016.0933.

7. Mohan Yu and Yajun Liu comment, "The firefly is the most efficient bioluminescent system for converting chemical energy into light with the extremely high luminescence efficiency." Mohan Yu and Yajun Liu, "A QM/MM

Study on the Initiation Reaction of Firefly Bioluminescence—Enzymatic Oxidation of Luciferin," *Molecules* 26, no. 14 (July 11, 2021): 4222, https://doi.org/10.3390/molecules26144222.

8. Hans E. Waldenmaier, Anderson G. Oliveira, and Cassius V. Stevani, "Thoughts on the Diversity of Convergent Evolution of Bioluminescence on Earth," *International Journal of Astrobiology* 11, no. 4 (2012): 335–343, https://doi.org/10.1017/S1473550412000146.

9. Anthony Fiorito III et al., "Spectral Fingerprint of Laser Emission from Rhodamine 6g Infused Male Indian Peafowl Tail Feathers," *Scientific Reports* 15, no. 20938 (2025), https://doi.org/10.1038/s41598-025-04039-8.

10. Kartik Aiyer, "How Extremophiles Push the Limits of Life," *American Society for Microbiology*, March 13, 2023, https://asm.org/articles/2023/march/how -extremophiles-push-the-limits-of-life .

11. Lynda Beladjal et al., "Life from the Ashes: Survival of Dry Bacterial Spores after Very High Temperature Exposure," *Extremophiles* 22 (2018): 751–759, https://doi.org/10.1007/s00792-018-1035-6.

CHAPTER 21: WILL THE REAL PRO-SCIENCE PARADIGM PLEASE STAND UP?

1. Richard Dawkins, *The Greatest Show on Earth: The Evidence for Evolution* (New York: Free Press, 2009), 419.

2. Paul Davies, *The Fifth Miracle: The Search for the Origin and Meaning of Life* (New York: Simon and Schuster, 1998), xvii.

3. Steven Vogel, *Cats' Paws and Catapults: Mechanical Worlds of Nature and People* (New York: W. W. Norton, 1998), 23.

4. Theodosius Dobzhansky, "Nothing in Biology Makes Sense Except in the Light of Evolution," *The American Biology Teacher* 35, no. 3 (1973): 125–129, https://doi.org/10.2307/4444260.

5. Philip Skell, "Why Do We Invoke Darwin?," *The Scientist*, August 28, 2005, https://www.the-scientist.com/why-do-we-invoke-darwin-48438.

6. Elizabeth Pennisi, "Haeckel's Embryos: Fraud Rediscovered," *Science* 277, no. 5331 (1997): 1435, doi: 10.1126/science.277.5331.1435a.

7. Casey Luskin, "What Do Modern Textbooks Really Say About Haeckel's Embryos?," *Discovery Institute*, March 27, 2007, https://www.discovery.org/a/3935/.

8. For more on the indispensable role of Christianity in the birth of science, see Nancy R. Pearcey and Charles B. Thaxton, *The Soul of Science: Christian Faith and Natural Philosophy* (Wheaton, IL: Crossway Books, 1994), Rodney Stark, *The Victory of Reason: How Christianity Led to Freedom, Capitalism, and Western Success* (New York: Random House, 2005), and Melissa Cain Travis, *Thinking God's Thoughts: Johannes Kepler and the Miracle of Cosmic Comprehensibility* (Moscow, ID: Roman Roads Press, 2022).

9. Paul Marks, "Float Like a Robot Butterfly," *New Scientist*, April 11, 2007, https://www.newscientist.com/article/mg19425996-300-float-like-a -robot-butterfly/.

10. Quoted in Jonathan Wells, "'In China We Can Criticize Darwin': Prelude," *Evolution News & Science Today*, April 16, 2014, https://evolutionnews.org /2014/04/in_china_we_can/.

11. Richard Dawkins, "Put Your Money on Evolution," *New York Times Review of Books*, April 9, 1989.

12. Richard Dawkins, *The Blind Watchmaker* (New York: Norton, 1986), 6.

CHAPTER 22: MY EXPERIENCE ADVOCATING INTELLIGENT DESIGN

1. Michael F. Ashby, *Material Selection in Mechanical Design*, 2nd ed. (Oxford, UK: Butterworth-Heinemann, 1999), 1.

2. Some evolutionists have put their hopes in what are known as Hox genes. These are master regulatory genes that impact where and when other genes are expressed during development. It is therefore proposed that mutations to Hox genes could cause major changes to body plans. But the experimental evidence stands squarely against such mutations. Stephen Meyer explains: "Despite the enthusiasm surrounding the field, evo-devo fails, and for an obvious reason: its main proposal, that early-acting developmental mutations can cause stably heritable, large-scale changes in animal body plans, contradicts the results of one hundred years of mutagenesis experiments.... The experiments of scientists such as Nüsslein-Volhard and Wieschaus have shown definitively that early-acting body-plan mutations invariably generate embryonic lethals—dead animals incapable of further evolution. The results of the experiments have generated the dilemma that... major changes are not viable; viable changes are not major. In neither case do the kinds of mutation that actually occur produce viable major changes of the kind necessary to build new body plans." Stephen C. Meyer, *Darwin's Doubt: The Explosive Origin of Animal Life and the Case for Intelligent Design* (New York: HarperOne, 2013), 314–315.

3. For a detailed analysis of the limitations of the evolutionary process as evidenced by a close study of computer algorithms designed to mimic biological evolution, see Robert J. Marks II, William A. Dembski, and Winston Ewert, *Introduction to Evolutionary Informatics* (Singapore: World Scientific Publishing, 2017).

4. Private email communication.

5. Jessica Shepherd and Steve Farrar, "Intelligent Design Creeps on to Courses," *Times Higher Education Supplement*, June 23, 2006, https://www.timeshighereducation.com/news/intelligent-design-creeps-on-to-courses/203867.article.

6. Stuart Burgess, "Against the Grain: 'There Are Strong Indications of Intelligent Design,'" *The Independent*, February 8, 2007, https://www.independent.co.uk/news/education/higher/against-the-grain-there-are-strong-indications-of-intelligent-design-435417.html.

7. Private email communication.

8. Email excerpt from Lewis Read, June 12, 2008.

9. "Peer-Reviewed Articles Supporting Intelligent Design," Discovery Institute Center for Science and Culture, https://www.discovery.org/id/peer-review/, accessed June 18, 2025.

10. Richard Dawkins, "Big Mistake," *Guardian*, December 27, 2006, https://www.theguardian.com/commentisfree/2006/dec/27/post845.

Figure Credits

Figure 1.1. Foot bones. "Sobo 1909 155." Image by Johannes Sobotta, 1909, Wikimedia Commons. Public domain.

Figure 1.2. Ligaments of foot. "Sobotta 1909 Fig. 228 Ligaments of the Foot, Superior/Lateral View." Image by Johannes Sobotta, 1909, and Dream_Studio3, AnatomyTool. org. CCA-SA license.

Figures 1.3a-d. Ankle movements. Images by Next Century Designs for Discovery Institute.

Figures 1.4a-c. Foot. Images by Next Century Designs for Discovery Institute.

Figure 1.5. Linkage fibula tibia. Image by Next Century Designs for Discovery Institute.

Figures 1.6. a-c. Ways of standing. Images by Next Century Designs for Discovery Institute.

Figure 1.7. Load paths. Image by Next Century Designs for Discovery Institute.

Figure 1.8. Bioinspired prosthetic foot. Photo by Joshua Carr. Used by permission.

Figures 1.9a-b. Roman block arch and foot arch. Images by Next Century Designs for Discovery Institute.

Figure 1.10. Construction jig. Images by Next Century Designs for Discovery Institute.

Figures 2.1a-b. Knee joints. Images by Next Century Designs for Discovery Institute.

Figures 2.2 a-b. Knee joint rotation. Diagrams by Stuart Burgess.

Figures 2.3a-b. Knee schematic. Images by Next Century Designs for Discovery Institute.

Figure 2.4a-b. Robotic knee joint. Photo by A. C. Etoundi. Used by permission.

Figure 3.1a. Wrist bones. "Sobo 1909 130." Image by Johannes Sobotta, 1909, Wikimedia Commons. Public domain.

Figure 3.1b. Wrist ligaments. "Sobo 1909 204." Image by Johannes Sobotta, 1909, Wikimedia Commons. Public domain.

Figures 3.2 a-b. Wrist rotation. Images by Next Century Designs for Discovery Institute.

Figures 3.3 a-c. Wrist adduction. Images by Next Century Designs for Discovery Institute.

Figures 3.4 a-b. Wrist flexion and rotation. Images by Next Century Designs for Discovery Institute.

Figures 3.5 a-b. Carpal tunnel. Images by Next Century Designs for Discovery Institute.

Figure 3.6. Load transfer function. Image by Next Century Designs for Discovery Institute.

Figure 4.1. Hand. Used by permission of the Florida Center for Instructional Technology.

Figures 4.2 a-c. Finger. Images by Next Century Designs for Discovery Institute.

Figures 4.3 a-d. Finger flexion and extension. Images by Next Century Designs for Discovery Institute.

Figure 4.4. Exoskeleton hand. Sketch generated by Stuart Burgess using ChatGPT.

Figure 5.1. Spinal column. "Vertebral Column, Lateral, Sobo 1909." Image by Johannes Sobotta, 1909, Wikimedia Commons. Public domain.

Figure 5.2. Tripod points pelvis. Images by Next Century Designs for Discovery Institute.

Figure 6.1. Mouth, nose, throat. "Medical Diagram of Anatomy of Nose, Mouth, Larynx, and Pharynx, with Annotations." Image by rob3000, Adobe Stock. Standard license.

Figure 7.1. Jaw. "File:913 Temporomandibular Joint." Image by Anatomy & Physiology, OpenStax, Rice University, 2013, Wikimedia Commons. CCA-3.0 license.

Figure 7.2. Chimp/human jaw side view. "Human and Chimpanzee Skull Biology and Anatomy." Image by kotjarko, Adobe Stock. Standard license.

Figure 7.3. Human/chimp mouth. Image by Next Century Designs for Discovery Institute.

Figure 8.1. Outer, middle, inner ear. Image by Next Century Designs for Discovery Institute.

Figure 8.2. Ear ligaments. Image by Next Century Designs for Discovery Institute.

Figure 8.3. Chain bones. Image by Next Century Designs for Discovery Institute.

Figure 8.4. Reptile ear. Image by Next Century Designs for Discovery Institute.

Figure 8.5. Clockface. Image by Next Century Designs for Discovery Institute.

Figure 8.6. Gears. "Gears and Cogs in Clockwork Watch Mechanism." Image by Photocreo Bednarek, Adobe Stock. Standard license.

Figure 9.1. Eye structure. "Structure of the Human Eye." Image by K3Star, Adobe Stock. Standard license.

Figure 9.2a. Extraocular muscles. "File:1107 The Extrinsic Eye Muscles Right Eye Lat." Image by CFCF, 2015, Wikimedia Commons. CCA-SA 4.0 license.

Figure 9.2b. Trochlea. "Eye Movements Lateral Rot." Image by Patrick J. Lynch, 2006, Wikimedia Commons. CCA 2.5 license.

Figure 9.3a. Human iris. "Close-up of the Iris of the Eye." Image by Chad Miller, November 13, 2005, Wikimedia Commons. Changed to grayscale. CCA-SA 2.0 license.

Figure 9.3b. Camera shutter. "Camera Shutter." Image by Cobalt, Adobe Stock. Standard license.

Figure 10.1. Human skin. "Human Anatomy, Skin and Hair Diagram, Integumentary System." Image by matoommi, Adobe Stock. Standard license.

Figure 11.1a-b. Birth. Images by Next Century Designs for Discovery Institute.

Figure 12.1. Human heart. "Human Heart Anatomy." Image by BlueRingMedia, Adobe Stock. Standard license.

Figure 13.1. Small intestine. "Cross Section of a Small Intestine." Image by Jazlyn G., 2020, Wikimedia Commons. CCA-SA 4.0 license.

Figure 13.2. Peristalsis motion. "Different Types of Peristalsis and Segmentation Motion." Image by vesvocrea, Adobe Stock. Standard license.

Figure 14.1. Muscle structure. "Illustration of Structure Skeletal Muscle Anatomy." Image by tigatelu, Adobe Stock. Standard license.

Figure 14.2. Vitruvian man. "File:0 The Vitruvian Man by Leonardo da Vinci." Image by Leonardo da Vinci, 1492, Wikimedia Commons. Public domain.

Figure 15.1. Nervous system. "Human Nervous System diagram." Image by Benevolent, 2014, Wikimedia Commons. CCA-SA 4.0 license.

Figure 15.2. Holes in skull. "Skull interior anatomy." Image by Patrick J. Lynch, 2006, Wikimedia Commons. CCA 2.5 Generic license.

Figure 15.3a. Nerve bundle. "Nerve Structure on White Background." Image by crevis, Adobe Stock. Standard license.

Figure 15.3b. Engineering cable. "Cable Wire on a White Background." Image by 3desc, Adobe Stock. Standard license.

Figure 16.1a-b. Satellite. Images by European Space Agency. Copyright ESA. Used by permission.

Figure 16.2. Double-action worm gearbox. Images by Stuart Burgess.

Figure 16.3. Efficiency equation. Stuart Burgess.

Figure 18.1. Human/animal arms. Image by Next Century Designs for Discovery Institute.

Figure 19.1. Spider silk. Image by Next Century Designs for Discovery Institute.

Figures 19.2a-f. Optimal structures. Image by Next Century Designs for Discovery Institute.

Figures 19.3a-h. Important linkage systems. Image by Next Century Designs for Discovery Institute.

Figure 19.4. Bacterial flagellum motor. Illustration by Joseph Condeelis/Light Productions for Discovery Institute.

Figure 19.5. Stuart Burgess diagram of predictions of how the human body has changed over time. Diagram recreated by Next Century Designs for Discovery Institute.

Acknowledgments

I would like to thank all those who have, either directly or indirectly, helped me write this book. There are many engineers, biologists, and chemists who have given me valuable advice. I cannot list them all, but I would like to particularly thank Steve Laufmann, Howard Glicksman, David Galloway, Casey Luskin, and Andy McIntosh. I have also been inspired by many talented people in the biomechanics community. I am especially grateful to Jonathan Witt for his expert editing. I would also like to thank my wife, Jocelyn, for her support and encouragement.

About the Author

Stuart Burgess, PhD, has held academic posts at the University of Cambridge, Liberty University, and the University of Bristol, where he served as the chair of the department of mechanical engineering. In the Rio, Tokyo, and Paris Olympics, he was the lead transmission designer for the British Olympic Cycling Team, helping them to gold medals in track cycling each time. His patented gearboxes appear on the four largest earth-observation satellites of the European Space Agency. He has published more than two hundred scientific papers on the science of design in engineering and biology. And he has received the following awards, fellowships, and appointments:

Awards

- 2019   James Clayton Prize
  (the UK's top mechanical engineering prize)
- 2017   Royal Society Summer Science Exhibition
  (Olympic bike design)
- 2017   IEOM Global Engineering Education Award
- 2013   Emerald Literati Award for Excellence
  (for publications)
- 2008   Wessex Institute Scientific Medal
  (for bioinspired design)
- 2001   IMechE Water Arbitration Prize
  (for engineering science)

- 1997 Turner's Bronze Medal (for mechanism design)
- 1993 Turner's Gold Medal (for spacecraft design)
- 1993 Mitutoyo Design Prize (for spacecraft design)
- 1986 Design Council Molins Prize (for mechanisms design)
- 1985 UK IMechE Queen's Silver Jubilee Second Prize

Fellowships

- 2021 Visiting Fellowship, Clare Hall College, Cambridge University
- 1994 Bye Fellowship, Selwyn College, Cambridge University

Appointments

- 2023 Guest Editor of the journal *Biomimetics*
- 2018 External Examiner of Engineering Tripos, Cambridge University
- 2011 External Examiner of Engineering, Exeter University
- 2006 Co-Editor of the *International Journal of Design & Nature and Ecodynamics*
- 2004 Chair of Department of Mechanical Engineering, Bristol University
- 1997 External Assessor of Engineering Research, Tokyo University

INDEX

www.ingramcontent.com/pod-product-compliance
Lightning Source LLC
Chambersburg PA
CBHW022101210326
41518CB00039B/351